LAI 风景园林与工业景观教席

高线启示
INSPIRATION HIGH LINE

围绕詹姆斯·科纳纽约代表作的国际学者评述

高线启示
INSPIRATION HIGH LINE

[德]乌多·维拉赫（Udo Weilacher） 编著

李梦一欣 黄琦 译

中国建筑工业出版社

目 录

Inhalt

中文版前言

　　2018 年 5 月 16 日，德国慕尼黑工业大学授予美国风景园林师詹姆斯·科纳（James Corner）荣誉博士学位。借此契机，《高线启示》原版正式发布，它也成为当时建筑学院风景园林与工业景观教席（德语：Lehrstuhl für Landschaftsarchitektur und industrielle Landschaft，LAI）^① 出版的系列丛书之一。

　　本书是多语汇编读本，共 27 篇短文。其中，20 篇为英语，7 篇为德语。全书汇集了由国际知名的风景园林领域的专家、学者，围绕高线公园撰写的一系列回顾、分析、评论与展望。文章篇幅短小，但视角独特、观点新颖、态度鲜明，蕴含着难能可贵的批判性思维。本书将有助于中国读者了解各国风景园林师对这一全球著名后工业景观设计作品的赞美与反思，从不同侧面揭示高线公园的成功所带来的巨大影响以及它背后不为人知的故事与情节。《高线启示》体现了风景园林师对新的城市自然的理解，对工业废弃地转型历程的诠释，对后工业景观设计原型和方法的回顾，以及对 21 世纪城市景观的展望与畅

---

① 2021 年建筑学院更名为工程与设计学院，风景园林与工业景观教席（LAI）更名为风景园林与景观转型教席（LAT）。

想。这些短文不同程度地揭示了高线公园项目在社会、生态、文化等方面给当代城市带来的持续变化。事实上，高线公园项目体现了詹姆斯·科纳将当代景观理论与实践相结合的探索，这一项目也是他拓展风景园林学科维度、并在文化层面实现"景观之想象"的重要媒介。

对中国读者来说，本书的出版将促使更多跨学科背景的专业人士全面了解高线公园项目。书中那些深刻且意味深长的评论或许会给我们带来更多启发和思考，而这一点，也成为我们着手翻译的初衷。

本书由慕尼黑工业大学风景园林学博士、北京建筑大学建筑与城市规划学院副教授李梦一欣和慕尼黑工业大学工程与设计学院研究员黄琦翻译完成。在整个过程中，我们得到了慕尼黑工业大学工程与设计学院，风景园林与景观转型教席负责人（德语：Lehrstuhl für Landschaftsarchitektur und Transformation，LAT）乌多·维拉赫教授（Prof. Dr. sc. ETH Zürich Udo Weilacher）以及意大利雷焦卡拉布里亚地中海大学、宾夕法尼亚大学瓦莱里奥·莫拉比托教授（Prof. Valerio Morabito）的大力支持，使本书以中、英、德三国语言出版。对此，我们致以最诚挚的感谢。最后，我们希望此书能得到全球范围内更广泛的支持与青睐，并使更多读者从中获益。

李梦一欣

2022 年 7 月

Chinese Preface

On 16 May 2018, the Technical University of Munich (TUM) awarded American landscape architect James Corner an honorary doctorate, which marks the official launch of *Inspiration High Line* as part of a series books published by the Chair of Landscape Architecture and Industrial Landscape (German: Lehrstuhl für Landschaftsarchitektur und industrielle Landschaft, LAI) at the School of Architecture at the time[1].

This book is a multilingual compilation of 27 different short readings, of which 20 are in English and the remaining 7 are in German. Written by the international renowned experts and scholars in the field of landscape architecture, the book brings together a wide range of reviews, analyses, comments and prospects on the High Line Park. Despite the short commentaries in the book, it still shows unique perspectives, novel ideas and explicit attitudes, implying valuable and precious critical thinking. By presenting

---

[1] The School of Architecture operating since 2021 as School of Engineering and Design and the Chair (LAI) as Chair for Landscape Architecture and Transformation (LAT).

Chinese readers with the praises and reflections of landscape architects from various countries on this world-renowned design work of post-industrial landscape, this book reveals not only the enormous impact caused by the success of the High Line Park in different respects, but also the untold stories and episodes behind it. *Inspiration High Line* demonstrates the understanding that landscape architects have of new urban nature, their interpretations of the transformation experienced by industrial derelict sites, their reviews on design paradigms and approaches to post-industrial landscape, as well as their expectations and visions for the urban landscape in the 21st century. Furthermore, these commentaries uncover the evolving changes that the High Line Park has brought about to the contemporary cities from such perspectives as society, ecology and culture. Admittedly, the High Line Park is what exemplifies the way James Corner explores the integration of contemporary landscape theory and practice. This typical project serves as an essential medium for his efforts to expand the dimensions of landscape architecture as a discipline and practice the Landscape Imagination on a cultural level.

For Chinese readers, the publication of *Inspiration High Line* will allow more and more interdisciplinary professionals to gain a full understanding of the park project. The insightful, thought-provoking commentaries included in this book are expected to provide further inspiration and prompt more reflection, which is our motivation to translate the book in the first place.

This book was translated by Mengyixin Li, PhD in landscape architecture, TUM and associate professor at the School of Architecture and Urban Planning, Beijing University of Civil Engineering and Architecture, and Qi Huang, researcher at the School of Engineering and Design, TUM. Throughout the process, we have been greatly supported by Prof. Dr. sc. ETH Zürich Udo Weilacher who is the chairman of Landscape Architecture and Transformation (German: Lehrstuhl für Landschaftsarchitektur und Transformation, LAT) at the School of Engineering and Design, TUM, and Prof. Valerio Morabito, Mediterranea University of Reggio Calabria and University of Pennsylvania. They have made it possible to publish this book in Chinese, English and German. We would like to express our mostly sincere thanks to them. Ultimately, we hope that the book will receive wider support and interest around the world, which will benefit many more readers.

Mengyixin Li

July 2022

驯服于纽约三楼的荒野

乌多·维拉赫

自 1905 年起，已有 37 位知名建筑师获得了慕尼黑工业大学授予的荣誉博士学位，他们分别在科学、技术和艺术方面作出了突出的贡献。2018 年，我校首次将这一学术荣誉颁发给一位风景园林师。詹姆斯·科纳，英国人，生于 1961 年，被誉为全球最具声望的风景园林师之一。他先后在英国和美国攻读风景园林专业，于 1986 年获得宾夕法尼亚大学硕士学位。之后，在国际顶尖的景观事务所积累了丰富的工作经验。从 1988 年开始，詹姆斯·科纳任教于美国费城的宾夕法尼亚大学。2000 年至 2012 年，担任建筑学院风景园林系主任。如今，他是国际上众多高校的客座教授，他的多项获奖作品曾屡屡在博物馆及美术馆展出。

詹姆斯·科纳是许多专业著作的作者、合著者及主编，在这些出版物中，他富有远见地探讨了人类世时期风景园林行业的职责，得到了专业领域的高度认可。《论当代景观建筑学的复兴》①（普林斯顿大学出版社，1999 年）是最重要的代表作

---

① 英文原著：Corner, J. *Recovering Landscape*：*Essays in Contemporary Landscape Architecture*. Princeton Architectural Press，1999. ——译者注

之一，书中汇集并拓展了后工业景观改建与振兴的重要见解。除了在学术上有所建树，科纳还在 1998 年创建了纽约 Field Operations 事务所。他在城市景观转型方面的示范性工作，吸引了业界和社会大众的广泛关注，比如纽约郊区 9 平方千米左右的清泉垃圾填埋场改建项目。面对这类高度复杂又需要长期建设的项目，事务所采取了以过程为导向的开发方法，具有开创性意义。

2009 年向公众开放的高线公园是当代风景园林设计的代表作，也是最热门的纽约地标。全长 2.5 千米的公园，曾经是 9 米高的废弃高架铁路，从曼哈顿西区的"肉类加工区"一直延伸到"地狱厨房"①。詹姆斯·科纳，既不是高线铁路荒野之美的发现者，也不是第一个把废弃铁路改建为公园的设计师。然而，他凭借精湛的设计能力让高线公园成为纽约最受欢迎的十大旅游景点之一，以及全球最著名的当代风景园林项目之一。

高线公园从何而来？1900 年前后，大约有 250 家屠宰场和肉类加工厂占据着哈德逊河沿岸的肉类加工厂街区。当时的地面货运铁轨与切尔西西区的街道交织在一起，货运列车和大卡车 24 小时不停地配送原材料和货品，让日常生活变得越来越危险。不久，第 10 大道就因为无数的致命车祸有了"死亡大道"之名。即使有"西区牛仔"骑行在每趟列车之前，举着红色警示旗，依然无济于事。因此，20 世纪初，人们大规模

---

① "地狱厨房"，其正式行政区名为克林顿（Clinton）。——译者注

改建了工业交通基础设施。1934 年，在 9 米高空架起了一条高架铁路，行驶的货运列车可以直接抵达工厂和西边仓库的楼上。

1950 年起，铁路运输量逐渐缩减。一方面，是因为工业生产从城市中迁出，另一方面，卡车送货的新交通系统得以实行。1960 年，人们首次拆除了一部分高架铁路。1980 年起，它被彻底废弃，尘封许久，纽约工业历史的"丰碑"从此变得锈迹斑斑。那些年，没有人去打扰野生又迷人的次生植被，它们就悄悄地生长在铁路道砟的碎石之间，城市荒野就这样隐匿在超级大都会的三楼。对于所有想要暂时避开公共视野的人来说，这个地方简直是天堂，他们也许是喜爱探险的孩童、恋爱中的青少年，也许是勇于突破的"异类"，又或者是寻求庇护的流浪者。

所以，当业主们为了提高私有地产的商业价值，打算剔除这个"眼中钉"时，反对的声音出现了。1999 年，两位来自曼哈顿的住户——罗伯特·哈蒙德和约书亚·戴维，极力主张保留高架铁路。废弃铁路的荒野之美，让他们认识到看似"生锈的怪物"背后承载的是城市历史的价值。于是，他们创办了"高线之友"协会。虽然，当时的市长鲁道夫·朱利安尼已经批准了第一期拆除许可。协会仍然在短时间内争取到来自艺术、文化和经济界众多人士的热烈响应与支持。最终，新组建的纽约市政府也选择支持他们的倡议。2002 年，时任市长迈克尔·布隆伯格签署了高架铁路转型为市民公园的决议。

次年，詹姆斯·科纳的 Field Operations 事务所与 Diller

Scofidio + Renfro 建筑事务所，以及荷兰植物设计师皮特·奥多夫共同合作，在高线公园改建的国际竞赛中胜出。团队的竞赛宣言是："保持它的简单、野生、沉静与舒缓"。设计参考了另一个案例，长约 4.5 千米的绿荫步道，在巴黎又叫作勒内·杜蒙绿色长廊。这个 1993 年开放的线性公园坐落在横贯巴黎第 12 街区的高架桥上，曾经也是一条 19 世纪建造的废弃铁路。詹姆斯·科纳表示："高线项目中，我们对场地有着深刻的分析和解读，其中两点最具代表性：一个是运输工程设施本身单调又自成一体的特点（例如：它的线性和重复性，与周围环境格格不入，钢铁材质的粗犷和混凝土的色调）；另一个是次生植物接管了停运后的工业结构，这种意外又迷人的景象，正如艺术家乔尔·斯特菲尔德在拍摄中捕捉到的凄美"[①]。

依照詹姆斯·科纳团队的设计方案，2006 年至 2014 年共分三个阶段，总耗资 2.6 亿美元，将线性废弃地转变为受欢迎的城市公园。它就像一根导火索，激发了周边建筑、商业和文化的繁荣，对城市荒野本身和周围的社会结构都产生了持续的影响。一旦作为向公众开放的日常生活空间，就会不可避免地改变废弃地的特征，高线公园也不例外。

Field Operations 事务所的目标是通过多功能、模块化搭建的肌理将货运高架铁路变成一个适宜步行的公园。因此，建造之前彻底清除了所有的道砟碎石、枕木、铁轨和野生植被。为了尽力还原先锋植物的荒野景象，植物设计师皮特·奥多夫搭

[①] 摘自詹姆斯·科纳：《与亨特相伴》，收录于克里斯托夫·林德纳，布莱恩·罗莎（编）：《解构高线——后工业都市主义和高架公园的兴起》，2017，p. 24.

配了草原草种、野生的多年生草本和典型的先锋树种（如白桦树）。铁轨和道砟碎石也被大面积地重新铺设。设计师还研发了 10 种不同的混凝土铺装组合——集户外座椅、照明和休息区为一体，搭建出便于使用的公园路面。预制的长条形混凝土铺装被精准地安装在支撑钢架上，铺装缝隙间穿插着不同的种植单元，构成像编织地毯一样的细致纹样。

高线公园是一个很好的范例，它展示了如何以专业的设计和发展理念应对城市基础设施的变化。在这里，自然性与人文性并不是相互对立的关系，而是融为一个全新的园林、建筑与技术相结合的有机体，在 9 米的高空串联着这座城市。但是，业内对高线公园的看法不一，不只有掌声，也存在一些质疑的声音，本书将从国际化的专业视角对此展开探讨。

来自 12 个不同国家的 27 位知名风景园林学教授，他们从各自的角度在文集中评述了这一热门的纽约废弃地转型项目，对于詹姆斯·科纳的风景园林设计方法和未来城市核心区基础设施改建的议题，贡献了富有启发性的学术讨论。

Gezähmte Wildnis im 3. Stock New Yorks

Udo Weilacher

Seit 1905 verlieh die Technische Universität München (TUM) die Würde eines Doktors ehrenhalber an 37 namhafte Architekten für herausragende wissenschaftliche, technische oder künstlerische Leistungen. 2018 wird erstmals einem Landschaftsarchitekten diese besondere akademische Ehrung zuteil. Der Brite James Corner, geboren 1961, zählt weltweit zu den renommiertesten Landschaftsarchitekten. Er studierte Landschaftsarchitektur in England und in den USA an der University of Pennsylvania, wo er 1986 seinen Masterabschluss erwarb. Nach erfolgreicher Mitarbeit in international führenden Landschaftsarchitekturbüros wurde er 1988 als Professor an die renommierte University of Pennsylvania berufen und leitete von 2000 bis 2012 als Dekan das Department of Landscape Architecture an der School of Design in Philadelphia. Corner ist Gastprofessor an zahlreichen internationalen Universitäten und stellte seine mehrfach ausgezeichneten Arbeiten bereits in vielen Museen aus.

Er ist Autor, Co-Autor und Herausgeber von vielen Fachpublikationen, die in Fachkreisen höchste Wertschätzung erfahren, weil sie weitsichtig das Wirken der Landschaftsarchitektur im Anthropozän untersuchen. Zu den erfolgreichsten Veröffentlichungen zählt *Recovering Landscape: Essays in Contemporary Landscape Architecture* (Princeton 1999), in welchem wichtige Erkenntnisse zum Umbau und zur Revitalisierung ehemals industriell genutzter Landschaften zusammengetragen und weiterentwickelt werden.

Corner publiziert jedoch nicht nur erfolgreich, sondern erregt mit seinem 1998 in New York gegründeten Büro „Field Operations" auch durch vorbildliche Arbeiten, bevorzugt in urbanen Transformationslandschaften, die Aufmerksamkeit der Fachwelt sowie der allgemeinen Öffentlichkeit. Hochkomplexe, langfristig angelegte Landschaftsprojekte, wie die Sanierung der etwa neun Quadratkilometer großen Mülldeponie *Fresh Kills* vor den Toren von New York City, gelten aufgrund ihres prozessorientierten Entwicklungsansatzes als besonders richtungsweisend und versprechen in Zukunft wichtige Erkenntnisse.

Zu einer Ikone aktueller Landschaftsarchitektur und zur populären Landmarke in New York City ist bereits der 2009 eröffnete High Line Park geworden. Die 2,5 Kilometer lange Anlage entstand in neun Metern Höhe auf einer ehemaligen Hochbahnstrecke, beginnt im Meatpacking District an der West Side von Manhattan und endet in Hell's Kitchen. James Corner ist weder der Entdecker der High Line noch der Erfinder von Parks auf stillgelegten Bahnstrecken, aber er sorgte mit seinem ausgezeichneten entwerferischen Können dafür, dass die High Line heute in New York zu den zehn beliebtesten Touristenattraktionen und weltweit zu den bekanntesten Projekten aktueller Landschaftsarchitektur zählt.

Wie entstand die High Line? Um 1900 dominierten noch etwa 250 Schlachthäuser und fleischverarbeitende Betriebe den Meatpacking District am Rande des Hudson River, und ebenerdige Eisenbahntrassen durchzogen die Straßen von West Chelsea. Güterzüge und Lastkraftwagen sorgten rund um die Uhr für die Verteilung von Rohstoffen und Waren, und das Straßenleben wurde immer gefährlicher. Die 10th Avenue war aufgrund zahlloser tödlicher Unfälle schon bald als „Death Avenue" bekannt. Selbst der Einsatz der „West Side Cowboys", die mit roten Warnflaggen jedem Zug auf Pferden voran ritten, konnten die Lage kaum verbessern. Deshalb entschloss man sich Anfang des 20. Jahrhunderts zu einem massiven Umbau der industriellen Verkehrsinfrastruktur und eröffnete 1934 eine auf neun Meter aufgeständerte, stählerne Hochbahntrasse, auf der die Güterzüge die oberen Stockwerke der zahlreichen Fabriken und Lagerhäuser der West Side direkt erreichen konnten.

Ab 1950 nahm der Schienenverkehr allmählich ab, weil die Industrieproduktion aus der Stadt verlagert wurde und sich ein neues Transportsystem durchsetzte: der Gütertransport per Lastkraftwagen. 1960 riss man erste Teile der Hochbahn ab und legte die Strecke 1980 endgültig still. Lange Jahre blieb das Monument der urbanen Industriegeschichte New Yorks ungenutzt und setzte Rost an. Auf dem Gleisschotter wuchs mit den Jahren nahezu völlig ungestört eine faszinierende, wilde Ruderalvegetation, urbane Wildnis im 3. Stock der Millionenmetropole und ein Paradies für all jene, die für eine Weile aus dem Blickfeld der reglementierten Stadtöffentlichkeit verschwinden wollten – abenteuerlustige Kids, verliebte Teenager, experimentierfreudige Aussteiger oder schutzsuchende Obdachlose.

Als Grundstückseigentümer im Interesse der Wertsteigerung ihrer Liegenschaften die Tilgung des „Schandflecks" aus der Stadt forderten, keimte Widerstand auf. 1999 machten sich zwei Anwohner aus Manhattan, Robert Hammond und Joshua David für den Erhalt der Hochbahnstrecke stark. Sie waren fasziniert von der wilden Schönheit der Bahnbrache, erkannten den stadthistorischen Wert des rostigen Monstrums und gründeten den Verein „Friends of the High Line". Obwohl Bürgermeister Rudolph Giuliani bereits erste Abrissgenehmigungen unterzeichnet hatte, gelang es dem Verein in kurzer Zeit so viele aktive Unterstützer aus Kunst, Kultur und lokaler Wirtschaft für ihre Idee zu gewinnen, dass schließlich auch die neue Stadtverwaltung von New York die Initiative unterstützte. 2002 unterzeichnete Bürgermeister Michael Bloomberg den Beschluss zur Umwandlung der Hochbahn in eine öffentliche Parkanlage.

Im Jahr danach gewannen den internationalen Wettbewerb zur

Umgestaltung der High Line James Corner mit „Field Operations"
in Zusammenarbeit mit Diller Scofidio + Renfro Architekten und
dem niederländischen Gartenarchitekten Piet Oudolf. Das Wettbe-
werbsmotto des Teams: „Keep it simple. Keep it wild. Keep it qui-
et. Keep it slow". Als gestalterisches Vorbild diente unter anderem
die etwa 4,5 Kilometer lange *Promenade planteé*, in Paris auch
als *Coulée verte René-Dumont* bekannt. Auch dieser lineare Park,
eröffnet 1993, entstand auf der Trasse einer stillgelegten Bahnlinie
des 19. Jahrhunderts, die auf Viadukten das 12. Arrondissement
von Paris durchquerte. „In the case of the High Line", erläutert
James Corner, „a very close reading was made of the site's history
and urban context. Two readings were particularly formative—
one was the singular, autonomous quality of the transportation
engineering infrastructure (its linearity and repetition, indifferent
to surrounding context, and its brash steel and concrete palette),
and the other was the surprising and charming effect of self-sown
vegetation taking over the postindustrial structure once the trains
had stopped running—a kind of melancholia captured beautifully
in earlier photographs made by the artist Joel Sternfeld." [1]

　　Zwischen 2006 und 2014 wurde nach den Plänen des Design-
teams um James Corner in drei Phasen für insgesamt etwa 260

---

① Corner, James: „Hunt's Haunts" in: Lindner, Christoph/ Rosa, Brian (Hrsg.):
Deconstructing the High Line. Postindustrial Urbanism and the Rise of the Elevated
Park. New Brunswick 2017; S.24

Millionen US Dollar ein lineares Stück urbaner Brache in einen populären Park verwandelt, der wie eine Zündschnur einen Bau-, Kommerzund Kultur-Boom in den angrenzenden Stadtquartieren auslöste. Das blieb für die gewachsene Sozialstruktur im Umfeld und für die urbane Wildnis nicht folgenlos. Wenn Brachflächen der Öffentlichkeit zugänglich gemacht und für die alltägliche Nutzung erschlossen werden, verändern sie zwangsläufig ihren Charakter – so auch die High Line.

Field Operations verwandelte die Hochbahntrasse in einen begehbaren Park mit einer multifunktionalen, modular aufgebauten Oberfläche, und im Zuge der Neugestaltung mussten Schotter, Schwellen und Schienen samt wilder Vegetation zunächst völlig entfernt werden. Um dem Erscheinungsbild der wilden Pioniervegetation möglichst nahe zu kommen, komponierte der Gartenarchitekt Piet Oudolf ein Pflanzschema aus Präriegräsern, Wildstauden und typischen Pioniergehölzen wie die Birke. Schienen und Schotter wurden wieder in weiten Teilen eingebaut, und um eine bequem nutzbare Oberfläche mit Sitzgelegenheiten, Beleuchtung und Aufenthaltsbereichen zu schaffen, entwickelten die Planer zehn unterschiedliche Plankenmodultypen aus Beton. Die Planken montierte man passgenau auf die stählerne Tragkonstruktion, und es entstand das Bild eines feinen Webteppichs, der in sein Gewebe verschiedene Lebensbereiche für die Vegetation integriert.

Der High Line Park kann als beispielgebendes Projekt gelten, weil es auf die Veränderung von urbaner Infrastruktur mit einem technisch ausgefeilten Entwicklungs- und Gestaltungskonzept reagiert. Natürlichkeit und Künstlichkeit stehen hier nicht im Gegensatz zueinander, sondern verbinden sich zu einem neuen landschafts-architektonisch-technischen Organismus, der die Stadt in neun Metern Höhe durchzieht. In Fachkreisen wird der High Line Park aber nicht nur mit Begeisterung, sondern teilweise auch mit gewisser Skepsis betrachtet. Und um die Sondierung dieses Spannungsverhältnisses aus internationaler fachlicher Perspektive geht es in dieser Publikation.

27 renommierte Professorinnen und Professoren der Landschaftsarchitektur aus 12 verschiedenen Nationen beleuchten in kurzen Essays aus ihrer persönlichen Sicht das populäre New Yorker Konversionsprojekt und leisten damit aufschlussreiche Beiträge zum fachlichen Diskurs über den zukünftigen Umbau innerstädtischer Infrastrukturen im Allgemeinen und James Corners landschaftsarchitektonischen Ansatz im Besonderen. •

拓展领域

托比约恩·安德森

20 世纪 80 年代的风景园林教学有时很困难——很难找到与之相关的当代书籍让学生阅读。的确，那时有很多关于自然科学、规划和城市社会学的书籍，但与建筑学著作的思想敏锐性相比，我们缺乏可借鉴的灵感，似乎风景园林不过是你能用眼睛看到的东西——一种缺乏足够理论的实践。

20 年过去了，21 世纪初许多重要的城市景观项目得以实现，风景园林教学开始走出困境，至少有更多当代的参考资料可供探讨。来自欧洲、美国甚至全世界厚厚的画册，各种网站、竞赛和众多奖项为我们直观地提供了铺天盖地的项目。我们有时却很难解释所有这些项目到底代表了什么，难道我们只是用过多的规划和科学手段换来了过量的景观意象？

因此，20 世纪 80 年代意味着对缺失的景观理论的不断渴求。21 世纪初呈现了大量的景观实践。在所有这些项目中，刊登最多的可能就是纽约的高线公园，甚至在它建成之前便已开始。

渴求是一种比满足感更强的驱动力，在我看来，这或许是詹姆斯·科纳努力成为作家的原因，也是他众多事业中最具影响力的一个。在缺乏理论和大量建成项目之间的这段时间里，他在写作、思考和演讲方面投入诸多，为我们提供理论叙述，用过激的言辞引起我们的不适感 ①，又以巧妙的结论进行安抚。有三本书给人留下了特别深刻的印象：第一本是《测量美国景观》（1996 年）②，科纳在航拍摄影师亚历克斯·麦克林的帮助下，重新审视了新的国土空间，撰写这样一本书是意义深远的，几乎是一场革命了。

　　第二本书《论当代景观建筑学的复兴》（1999 年），是科纳编著的一部论文集。其中包含 16 篇不同的文章，它们共同为风景园林行业提供了有力辩护和行动方法。书中的内容超越了图像学，提供了一种破解习语之谜的方式。科纳的文章引用充分，他曾用托马斯·斯特尔那斯·艾略特的诗作为开场白，使风景园林与罗兰·巴特、米歇尔·福柯、马塞尔·杜尚，甚至马丁·海德格尔相关联。科纳对风景园林的再定义使我们所处的行业一跃成为"重要文化活动"③ 的一部分。

　　第三本书《景观都市主义》（2006 年）④，是由查尔斯·瓦尔德海姆编著的另一部论文集，共 14 篇文章，其中的 1 篇由科纳撰写，文章同样充满攻击性。他似乎在说，既然建筑师占

---

① 詹姆斯·科纳以尖锐的批判性和启发式的写作方式在当代景观研究领域独树一帜。——译者注

② 英文原著：Corner, J., MacLean, S. A., Cosgrove, D. *Taking Measures Across the American Landscape*. Yale University Press, 1996. ——译者注

③ "Critical cultural activity" 出自《论当代景观建筑学的复兴》一书中詹姆斯·科纳的英文原文 "Recovering Landscape as a Critical Cultural Practice", pp. 1–26. ——译者注

④ 英文原著：Waldheim, C. *The Landscape Urbanism Reader*. Princeton Architectural Press, 2006. ——译者注

用了我们的景观概念，并将其简化为单纯的背景，我们就必须
阐明他们只理解了其复杂内涵的一小部分。我们应将景观引入
城市，证明景观属于乡村的观点早已过时。景观都市主义作为
一种态度具有很大的影响力，尽管这一概念在被视为一种方法
时遇到了一些定义上的问题。对许多建筑师而言，景观和都市
主义是难以相融的，但现在它们已合二为一。

Expanding the Field

Thorbjörn Andersson

Teaching landscape architecture in the 1980s was sometimes difficult — it was hard to find books that were relevant and contemporary to give the students. It is true, there were books available on the natural sciences, on planning, and on urban sociology, but compared to the intellectual acuity of architectural writing of the time, we were short on inspiration to draw upon. It was as if landscape architecture was no more than what you could see with your eyes — a practice lacking sufficient theory.

Two decades passed, and the beginning of this century saw a great number of significant landscape projects in the urban realm being realized. Teaching landscape architecture became somewhat simpler, at least there were far more contemporary references to discuss. Brick-thick picture books, websites, competitions, and numerous awards supplied us with an avalanche of photogenic projects, not only from Europe and the USA, but from all over the world. Occasionally we had a hard time interpreting what all these projects stood for. Had we just exchanged an overdose of planning and science for an overdose of imagery?

Thus, the 1980s meant a constant thirst for absent theory. On the other hand, the early 2000s displayed an abundance of practice. Of all the projects, probably the most published was the High Line in New York, beginning before it was even built.

Thirst is a stronger driving force than satisfaction, which is probably why I rate James Corner's endeavors as a writer as the most influential of his many undertakings. During the time between the period without theory and the period with abundant built projects, he took action in writing, thinking, and presenting aspects that supplied us with theory, provoked us with insults, and soothed us with clever conclusions. Three books made an especially deep impression: In the first, *Taking Measures Across the American Landscape* (1996), Corner examined new territories afresh with the help of aerial photographer Alex MacLean. Writing such a book was bold, almost a form of crusade and an act of reverse colonization.

The second book is an anthology that Corner edited, entitled *Recovering Landscape* (1999) . It contains sixteen disparate essays, that together served as a powerful defense for the profession and a method for action. The book was a way to crack open the riddle of the idiom itself, to go beyond the iconography. Corner's own text is full of references. He had opened *Measures* ... with a poem by T.S. Eliot. Here, landscape architecture was related to

Roland Barthes, Michel Foucault, Marcel Duchamp and even Martin Heidegger. We suddenly became part of a "critical cultural activity", to use Corner's own re-definition of landscape architecture.

The third book, *The Landscape Urbanism Reader* (2006), is another anthology, edited by Charles Waldheim, with fourteen essays, with one by Corner that serves as an offensive maneuver. He seems to be saying that since architects have appropriated our concept of landscape and reduced it to mere background, we must make it clear that they understand only a fraction of its complexity. We should bring landscape into the city, and prove obsolete the belief that landscape belongs in the countryside. Landscape urbanism came to be influential as an attitude, although the concept ran into some definition problems when seen as a technique. Landscape and urbanism had been oil and water to many architects; now they were unified. •

高线启示：荷兰案例

阿德里·范·登·布林克

2009 年，自高线公园开放以来，它一直被誉为废弃基础设施转向高品质城市环境的典范。它是在曼哈顿西区旧铁路线上构建的公共景观，根植于可持续增长这一城市议题。从一开始，它就为想要逃离繁忙都市生活的当地人形成了一片绿洲，同时作为一处广受欢迎的旅游景点，为塑造当代城市形象起着显著作用。荷兰园艺师皮特·奥多夫和詹姆斯·科纳合作，沿漫步道创造了不同的空间领域和生物栖息地，为高线公园的成功做出了重要贡献。

除了赢得好评，高线公园对利润的关注超过了对地球和人的关注，也因此遭受批评，导致诸如绅士化和社会排斥现象。这些影响难以衡量，但应与其他解决方案所产生的影响进行权衡。事实上，人们也许更加反对公园与周边环境的脱节，因为整个项目更像是一个逃离城市生活的避难所，而非与之融合的组成部分。从中至少可以吸取的经验是，类似的城市改造项目应有助于创造不同类型的联系：城市各部分之间的联系，公园不同使用者之间的联系，以及 3 个"P"① 之间的联系，即可持

① 代表首字母为 P 的 3 个英文单词：People, Planet, Profit。——译者注

续发展三角中的人、地球和利润。

　　几位荷兰的建筑师和风景园林师为高线公园思想的传递做出了巨大贡献。例如，荷兰建筑事务所 MVRDV 和景观设计师本·奎珀斯设计的首尔 Seoullo 7017 项目。在这一案例中，位于首尔市中心的高速公路立交桥由于不再符合安全标准而被改造为一处"空中花园"，它将成为一条服务当地居民的生命线，并作为城市可持续发展的标志。这条穿过主火车站的高空步道连接着周边各个街区，形似一条章鱼。该项目不仅是一条每月可承载 200 万游客的空中走廊——附近有咖啡馆、商店和办公楼——也是城市中一个可以让人逗留、相遇和散步的地方。

　　荷兰也有其他类似的案例。由 ZUS 建筑事务所设计的 Luchtsingel 是一座 400 米长的人行天桥，它重新连接了鹿特丹市中心的三个区域。这一人行天桥已作为城市更新和经济发展的催化剂发挥着自身作用。在被称为"永恒的瞬间"设计理念的基础上，项目整合了一系列绿色公共空间，包括欧洲第一个都市屋顶农场。事实上，它诠释了一个完整的众筹项目是如何嵌入当地城市社会生活的。由景观事务所 OKRA 设计的东部铁路公园，位于荷兰乌得勒支市郊，连接着市中心和乡村地区①。铁轨作为历史的痕迹得以保留。在市民的积极参与下，这一"低线"公园②得以开发，包括多条步行小径、一段长途骑行

①　从功能上，公园起到了绿带的作用，详见项目官网。——译者注
②　新的东部铁路公园被设计师视作"非官方的高线公园"，没有明显架高的上层空间，因而被解读为"低线"公园。——译者注

路线和若干城市农场。整个项目可能不足以吸引游客与建筑杂志，但对于经常使用铁轨公园的当地人来说却颇具趣味。

　　这些案例显示了在詹姆斯·科纳先锋设计的影响下，荷兰专业人士是如何接受具体挑战的。在所谓"荷兰方法"的基础上，他们将土地利用中单一的线性基础设施转变为绿色宜居区域，并以多种方式促进城市可持续发展。[①]

---

① 更多信息详见 MVRDV（Seoullo 7017）、ZUS（Luchtsingel）和 OKRA（Oosterspoorbaanpark）官网。

Inspired by the High Line: Dutch Examples

Adri van den Brink

Since its opening in 2009, the High Line has been acclaimed as a showcase of how unused derelict infrastructure can be transformed into a high-quality urban environment. The public landscape that was created on the former railway line in West Manhattan is firmly rooted in the city's sustainable growth agenda. From the very start it has formed an oasis for locals who want to escape from busy city life, and also a popular tourist attraction that plays a prominent role in the contemporary image of the city. The different spheres and biotopes that the Dutch garden designer Piet Oudolf created along the promenade in collaboration with James Corner have been a major contribution to the High Line's success.

Apart from the praise it receives, the High Line has also been criticised for having a stronger focus on profit than on planet and people, resulting in, for example, gentrification and social exclusion. Such effects are difficult to measure but should be weighed against the effects of alternative solutions. A more serious objection may be that the park is rather disconnected from the surrounding urban environment; it is more a refuge from city life than an integrated part of it. One of the lessons that might be drawn from the project is that urban transformation of this kind should contribute to creating different kinds of connections: connections between the various parts of the city, between different users of the park and between the three P's: people, planet and profit of the sustainability development triangle.

Several Dutch architects and landscape architects have made a substantial contribution to cultivating the seeds that were planted by the High Line. One example is the *Seoullo 7017* project in Seoul by MVRDV and Ben Kuipers. In this case a motorway flyover in the heart of the city that no longer met safety standards was converted into a "skygarden" that would become a lifeline for local inhabitants and an icon of sustainable urban development. Like an octopus, the elevated promenades connect different neighbourhoods across the main railway station. With two million visitors per month it is not only a passage but—with nearby cafes, shops and office buildings—also a place to stay, to meet and to take a walk.

Other examples can be found in the Netherlands. The *Luchtsingel*, designed by ZUS, is a 400-m-long pedestrian bridge that reconnects three downtown districts in Rotterdam. The Luchtsingel, too, has proven itselfas a catalyst for urban renewal and economic development. Based on what is called "permanent temporality", it combines a series of green public spaces, including Europe's first urban farming roof. The fact that the project is completely crowd-funded illustrates how it is embedded in urban social life. The *Oosterspoorbaanpark*, designed by OKRA, is situated in the outskirts of Utrecht, connecting the downtown area with the countryside. Rails have been maintained as a historical reference. This "low line" park, developed with strong civic participation, contains footpaths, a section of a long-distance cycling route and urban farming sites. Perhaps not so interesting for tourists and architecture magazines, but all the more so for the locals who regularly use the park.

These examples show how Dutch professionals have taken up the challenge set by James Corner's pioneering design. Building on what may be called the "Dutch approach", they have transformed single land-use infrastructure lines into green liveable areas that contribute to urban sustainable development in multiple ways. •

交错

保罗·比尔吉

　　几年前，在威尼斯建筑大学的一门研究生课程中，我们让学生重读了路德维柯·阿里奥斯托的史诗《疯狂的罗兰》[①]，并为摩德纳省[②]的一个景观项目寻找灵感，我们认为，这可以为风景园林教育中常见的文献法提供一种补充。学生对课程的兴趣超出了所有人的预期，以至于不得不首次限制学生数量，将课程的注册人数减少到 50 人。

　　在阅读阿里奥斯托的作品时，学生获得了许多意想不到的灵感。16 世纪，世界充满着诗意的幻想与美丽的梦想（同时也不失新鲜感和讽刺意味）。此外，"交错"被认为是最具吸引力的概念。它是一种文学技巧，将几个故事同时交织在一个较大的叙事体中，混合不同素材，赋予其既含蓄又明确的文学内涵。通过这一手法，叙事序列被打断、分离，随后与其他叙事序列重新组合。

　　自科纳的高线公园项目出版以来，首先吸引观察者的设计重点便是现在这一非常著名的模式：一种铺装与种植的互动，

---

① 英文原著：Ariosto, L. *Orlando furioso*. Penguin Classics，1977. ——译者注
② 意大利北部城市。——译者注

实与空，刚与柔，石块与草地贯穿着整个线性公园。这是一个反复出现的主题，来来往往，时隐时现，自我再生，似一种邀请，融入相互交织的故事之中：以其最迷人的形式"交错"在一起。

在大学的风景园林教学中，工作室中的桌面研究反映了大多数学生对当下景观设计的兴趣。学生桌上的书籍，参观和浏览过的地方反映了其自身关注的焦点，他们时刻在寻找灵感，并总是在当代参考文献中为将来的设想寻找想法和主题。

因此，在设计过程中，学生经常会尝试将高线公园的概念重新解读为"交错"，然而结果并非总是像我们所期望的那样，呈现出一个有远见的设计方案。毕竟，复制品通常无法媲美艺术家的原作，然而在令人着迷的重新解读的过程中，仍有可能隐藏着创造的潜力。

去年，当我漫步在波哥大①市中心一个优雅的街区时，出乎意料地发现了一个花园，它和纽约高线公园的一部分高度相似：铺装上相互交错的线条从地面延伸至折叠长椅，狭窄的条形缝隙中，植被尝试着获得一些生长空间，然而却是徒劳的。我很难说这一设计是否受到高线公园项目的启发，这就如同其他未解之谜，还是留给未来的历史学家来解答吧。

---

① 哥伦比亚首都。——译者注

Entrelacement

Paolo Bürgi

A few years ago, during a masters-level course at the IUAV in Venice, we asked our students to re-read Ludovico Ariosto's epic poem *Orlando furioso* and find inspiration for a landscape project that was to be located in the Province of Modena, as a remedy we thought, to more literal strategies we often find in landscape architecture education. The interest from the students surpassed all expectations. For the first time, a limit to the number of students had to be introduced, whittling the number of enrolled students down to fifty.

While reading Ariosto, the sixteenth-century world with a rich tapestry of poetic fantasy and beautiful dreams (that have meanwhile lost nothing of their freshness and irony) many unexpected inspirations came to the students'. Out of these, one idea was agreed as being the most fascinating: that of *"entrelacement"*, a literary technique in which several stories are simultaneously interlaced in one larger narrative; mixing heterogeneous materials to give implicit as well as explicit meanings. As such, the narrative sequences are interrupted, separated and recombined with other narrative sequences later on.

Since the very early beginning of Corner's High Line Park project publications, the first design that captured the observer is for sure the now so famous pattern: a game of paving and planting, full and empty, hard and soft, rock and grass that traverse this linear park. It is a recurring theme; coming and going, appearing and disappearing, regenerating itself like an invitation to merge into interlaced stories: *entrelacement* in its most intriguing form.

In teaching landscape architecture at university, desk critics in studio are the mirror of students' most current interests in landscape architecture. The books on their tables and the places they visit and navigate through reflect where they focus their attention, searching for inspiration on the move. It is mostly in contemporary references that they search for ideas and topics for tomorrow's visions.

So in the students' design process, an attempt to reformulate the idea of the High Line as *entrelacement* is often a recurring theme, but the results are not always what we expect to be a visionary proposal. After all, copies usually are inferior to the artist's original, yet there may still be creative potential hidden in a ravishing reinterpretation.

Last year, while walking through an elegant neighborhood in downtown Bogota, I surprisingly and unexpectedly found a garden that was clearly a close relative of part of the High Line in New York: interlaced lines in the paving that merged out from the ground into a folded bench, while stripes of a shy vegetation vainly tried to gain some space within the narrow bounds. Whether this was indeed inspired by the High Line project, I cannot say. This, like many other questions, is one left for future historians. •

预感

米歇尔·德斯维涅

2000 年左右，我住在波士顿。詹姆斯·科纳和我偶尔会参加我们各自的大学——哈佛大学和宾夕法尼亚大学工作室之间的评论交流会。有一次我去费城，在他狭小的学校办公室里，詹姆斯向我展示了他当时为纽约一个看似不可能实现的项目所绘制的一些图纸。

我被它震惊了，且有着充分的理由。首先，我惊讶于项目的表现技法。这之前，詹姆斯就让我印象颇深，就像他给每个人留下的印象一样，他在《测量美国景观》中所呈现的地图，其视角的精准性和现实性在当时都是前所未有的。

但最重要的是，我被这个项目本身震惊了，不用说，正是高线公园。当然，我脑海中浮现出了一个虽种植一新却已过时的巴黎林荫道。但当时，已经可以从他的图纸中看出，在新旧结构之间所体现的一种令人难以置信的精致感与现代性。詹姆斯满怀期待，自豪地向我展示了著名的锥形混凝土单元，它被用来确保植物之间的过渡。

20 年过去了，我无数次漫步在清晨的高线公园，这个转瞬即逝的回忆依旧萦绕在心头。我惊讶地发现，这里的景观和詹姆斯图纸里预期的画面重合了。

Premonitions

Michel Desvigne

At the turn of the century, I was living in Boston. Occasionally James Corner and I would participate in studio review exchanges between our respective universities, Harvard and Pennsylvania. During one of my visits to Philadelphia, in his small university office, James showed me a number of drawings he was working on for a quite improbable project in New York.

I was stunned by it, and for good reason. To begin with, by the representational technique alone. James had greatly impressed me before, as he had everyone, with his maps in *Taking Measures Across the American Landscape*, but the precision and realism of these perspectives was of a quality completely unprecedented at the time.

Above all though, I was stunned by the project itself; which was, needless to say, the High Line. Of course, a newly planted but antiquated Parisian promenade came to mind. But here, visible already in his drawings, the play between the new development and the existing structure was of an unbelievable refinement and modernity. With much anticipation, James proudly showed me the famous tapering concrete section ensuring the transition between the plantings.

Twenty years and numerous morning walks along the High Line later, this fleeting vision continues to haunt me. I can only see this landscape superimposed over this sudden image of anticipation. •

荒草音律

约格·德特马

　　高线公园的故事开始于社会影响力广泛的公民参与，最先引发公众热议的其实是废弃铁轨上的次生自然①。后来才是我们熟悉的部分，即"高线之友协会"与纽约市政府共同合作，斥巨资打造了詹姆斯·科纳主持设计的高线公园。回顾过去 20 年来最著名、最特别且造价最高的风景园林项目，高线公园无疑是其中之一。这个公园也深受纽约市民和游客的喜爱，从这方面来说，它是一个成功的项目。不过，受欢迎的同时，它也给曼哈顿西区带来了严重的绅士化问题。

　　但是，高线公园的成功与绅士化并不是本文的核心。我希望探讨的其实是蕴含在设计之中的自然景象，以及自然在此扮演着怎样的角色。虽然肆意生长了 25 年的野生植物已经在高线铁轨的整修中被全部清除，但却赋予了改建后的植物配置灵感。担任植物设计师的皮特·奥多夫将场地原有的各类野生植物与新添加的物种融为一体。园艺优化后的整体状态能令人

---

① 次生自然（Sukzessionsnatur），与没有受到人类活动所干扰的"原始自然"相对应。次生演替（Sukzession），生物学上的植物群落由低级到高级，从简单到复杂的自然演变现象。——译者注

联想起"荒野"的景象，仿佛被重新调整的"荒草"的音律，营造出一幅宛如野生的城市自然景象。

对此我想继续追问：上述"自然"的形式到底象征着什么？欧洲启蒙运动之后，公园与花园曾长期象征着远离城市喧嚣的理想化风景，不仅被视为城市居民向往的地方，也代表着与城市截然不同的世界。之所以形成这种看法，是因为那时的城市环境与园林景观之间，存在着明显的差别。如今的城市自然是自发的、野生的，这种"自然"即使没有园艺培育也高度适应了城市环境。它在象征层面上代表着城市的生活条件和水平，并不是以往风景园林里能让人心生向往的自然，在某种程度上甚至令人避之不及。城市自然一直像是对城市秩序的威胁，是萧条的表现，例如收缩城市^①里野草丛生的衰败景象。

后现代主义时期，人们开始探讨自然在城市中扮演的多种角色。这渐渐地改变了城市和风景之间的对立关系，二者越来越融为一体。虽然已有许多针对城市原生自然的设计尝试，但其中大多数都没有被理解为"令人向往的地方"，只是在功能上提供了艺术与人文的场地。例如柏林和鲁尔地区的"次生公园"^②，人们将荒野自然保留为公园要素，通过小部分节点的景观设计，塑造了大面积的"城市自然"，更确切地说是"工

---

① 收缩城市（Schrumpfende Städte），多数指失业人口增多、人口外流，房屋空置率高的传统工业城市，比如美国底特律，德国波鸿等城市。自然重新占领了被废弃的建筑设施和场地，出现了无人问津、荒草丛生的衰败景象。——译者注
② 次生公园（Ruderalpark），原生植被和环境已经被破坏，形成了新的生长环境，比如战后瓦砾、旧铁路砾石、工业荒地或者道路铺装缝隙等。——译者注

业自然"主题。最著名的代表当属杜伊斯堡北景观公园。不过，这些案例都是高度保留工业文化遗迹的地方，是既富有人文气息又真正野生的城市自然，是粗放式管理维护的空间。

　　此前，从未有过像高线公园这样的项目，不仅仿造野生般的城市自然，还采用园艺技术精细化地养护这一景致，至少在这样的尺度和意义上是史无前例的。倘若荒野般的城市自然是园林设计模仿出来的，是精心种植的，这在象征层面上又意味着什么呢？原本会对城市秩序构成威胁的次生演替，被高线公园以美的方式呈现了出来。从欧洲的角度来说，这代表着一个可以被人类掌控的城市自然；一个被营造成园林般令人向往的地方，让人们可以无忧无虑地欣赏城市自然之美；这象征着城市与自然在美学上的琴瑟和鸣。

Tuning the Weed

Jörg Dettmar

Der High Line Park geht zurück auf starkes bürgerschaftliches Engagement. Ein Auslöser für dieses Interesse war die Sukzessionsnatur auf den ehemaligen Gleisen. Die weitere Geschichte ist bekannt; der Park wurde mit gewaltigen finanziellen Mitteln in der Zusammenarbeit der Stadt NYC und den „Friends of the High Line" auf der Basis des Entwurfs von James Corner realisiert. Es ist sicher eines der interessantesten, teuersten und bekanntesten landschaftsarchitektonischen Projekte der letzten 20 Jahre. Insofern ist der Park ein großer Erfolg und bei vielen New Yorkern und Touristen sehr beliebt. Er ist allerdings auch Auslöser für eine massive Gentrifizierung in West Manhattan.

Dies soll hier aber nicht im Fokus stehen, sondern mich interessiert die Frage nach den Naturbildern und der Rolle von Natur, die hinter der Gestaltung stehen. Die spontane Vegetation der fünfundzwanzigjährigen Verwilderung wurde bei der Sanierung der Gleistrasse entfernt, diente aber bei der Neubepflanzung als Inspiration. Piet Oudolfals ausführender Vegetationsplaner hat verschiedene vorher hier vorkommende Arten in die Neubepflanzung integriert und insgesamt ein Vegetationsbestand geschaffen, der an die „Wildnis" erinnert aber gärtnerisch optimiert ist. Neugebaut hat man ein Bild von wilder Stadtnatur oder anders ausgedrückt das „Unkraut" wurde getunt.

Interessant ist daran für mich die Frage was diese Form der „Natur" nun eigentlich symbolisiert? In Europa symbolisierten Parks und Gärten seit der Aufklärung lange eine ideale Landschaft außerhalb der Stadt als Sehnsuchtsort des Städters und Gegenwelt zur Stadt. Damit verbunden war die klare Trennung von Stadt und Landschaft. Stadtnatur als eigentliche, wilde, nicht gärtnerisch angelegte und gut an die Stadt angepasste Natur ist auf der symbolischen Ebene Ausdruck der städtischen Lebensbedingungen und in gewisser Weise der Gegensatz von sehnsuchtsbeladener Natur der Landschaft. Sie ist immer auch bedrohlich für die städtische Ordnung und Ausdruck des Zerfalls, dieses Phänomen kann man in schrumpfenden Städten erleben.

Spätestens in der Postmoderne und seit den Diskursen über die Rolle der Natur in der Stadt hat sich etwas verändert in dem Verhältnis von Stadt und Landschaft, der Gegensatz wurde mehr und mehr zu einer hybriden Gesamtheit. Es gibt viele Versuche originär urbaner Grünflächengestaltung, die aber meist nicht als „Sehnsuchtsorte" sondern mehr als Kunst- oder Kulturorte funktionieren. Die „Ruderalparks" in Berlin und besonders im Ruhrgebiet bei denen man die spontane Natur als Element eines Parks erhalten und teilweise auch landschaftsarchitektonisch inszeniert hat, thematisierten großflächig die Stadt- beziehungsweise „Industrienatur". Bekanntestes Beispiel ist wahrscheinlich der *Landschaftspark Duisburg-Nord*. Allerdings sind dies Reservate und extensiv kontrollierte Räume einer kulturell aufgeladenen aber immer noch authentischen, weil originären Stadtnatur.

Der Nachbau der wilden Stadtnatur, das gärtnerisch kontrollierte und intensiv gepflegte Abbild – wie bei der High Line – hat es bislang zumindest in der Dimension und mit dieser Bedeutung nicht gegeben. Was bedeutet es nun also auf symbolischer Ebene wenn wilde Stadtnatur gärtnerisch imitiert und mit extrem hohem Aufwand kultiviert wird? Aus europäischer Sicht würde ich sagen, dies ästhetisiert und inszeniert die von der Sukzession eigentlich ausgehende Bedrohung städtischer Ordnung, es signalisiert eine beherrschbare Stadtnatur und eignet sich damit als Sehnsuchtsraum für deren ungefährlichen Genuss. Sie symbolisiert ästhetische Funktionalharmonie zwischen Stadt und Natur. •

林荫道的新范式

索尼娅·邓佩尔曼

　　通过高线公园项目,詹姆斯·科纳和他的 Field Operations 事务所设计了一处地标性景观,为城市林荫道提供了一种新范式。在纽约,21 世纪的高线公园可被认为是类似于 19 世纪中央公园的一条步行道,它甚至可以与弗雷德·劳·奥姆斯特德和卡尔弗特·沃克斯设计的极具标志性的中央公园相媲美。同样地,高线公园也吸引了美国大众、市政官员以及国内外城市与景观设计师的注意。它是纽约市最大的旅游景点之一——虽然只是大致上模仿了巴黎的林荫大道——但其他城市早已纷纷效仿,例如,芝加哥已实施的布卢明代尔 3 英里步道,费城开始建设的雷丁高架铁路公园,以及圣路易斯市按照高线公园模式所构想的栈桥公园。

　　高线公园在美学策略、设计语汇、图示、象征意义及其对周围社区的经济影响等方面都与 19 世纪的城市公园发展类似。例如,一旦将原先的高架货运线改造成线性公园的计划被推动,周边地区就开始增值,进一步促进和加快了原先肉类加工区的绅士化。如果说中央公园被设计为一个僻静的、田园般、风景如画的世外桃源,那么高线公园的设计师则为人们提供了一条绿色长廊，使游人既能脱离街道，又能成为其中的一部分。中央公园被19世纪的设计师设想为一片绿洲，周围密集的城市结构被掩盖其中。与此相反，高线公园的特征和布局却蕴含着人们对城市的美好体验，通过各种带座椅的观景平台以及框景设施可以直接俯瞰城市街道生活、街谷，并唤起一种崇高的美学效果，这与21世纪初许多作家对纽约街道被高楼大厦包围时的描述相类似。

高线公园的大部分植物由城市腹地的常见树种组成，它们模仿了在荒废的轨道之间自发形成的植被，为建筑物之间和屋顶上的远景提供了一种崇高的氛围和多彩的风景框架。在铺装和城市家具的设计中，高线公园使人回想起旧轨道结构的运输功能，这一项目在美化场地工业历史的同时，使一切变得可见。

通过这些构筑物，这条高耸的长廊既与城市分开，又与城市相连。这项设计使一条废弃的工业铁路变成一种便利设施。它既突出又掩盖了该地区以前的工业特征，并将其发展为一个高端旅游目的地和富裕居民的住宅环境。在这个意义上，高线公园的设计可能比19世纪的中央公园更加简洁明了。通过回顾场地的历史，对原货运线路进行的适应性设计，为改善城市景观开创了先例。

A New Paradigm for the Promenade

Sonja Dümpelmann

With the project of the High Line, James Corner and his firm Field Operations have provided a landscape of landmark quality and a new paradigm for the promenade. In New York City, the High Line could be considered the twenty-first-century equivalent of the nineteenth-century Mall in *Central Park*, or even of the entire iconic *Central Park* designed by Frederick Law Olmsted and Calvert Vaux. Like Central Park, the High Line has attracted both the attention of a large public and of city officials, urban and landscape designers in the United States and abroad. It ranks among the biggest tourist attractions in the city and—while itself modeled loosely on the *Promenade Plantée* in Paris—has been and is being emulated by other cities. For example, Chicago has implemented the three-mile long *Bloomingdale Trail*; Philadelphia has begun to build the *Reading Viaduct Rail Park*; and St. Louis has envisioned *The Trestle* following the High Line model.

The High Line also parallels nineteenth-century public urban park developments in terms of its aesthetic strategies, design vocabulary, iconography, symbolism, and its economic effects on the surrounding neighborhoods. For example, once the plans to turn the former elevated freight line into a linear park gained traction, the surrounding area began increasing in value, facilitating and further accelerating the gentrification of the former meatpacking district. Yet, if Central Park was designed as a secluded, introverted pastoral and picturesque antidote to the city, with the High Line the designers have provided a green promenade that enables visitors to both detach themselves from the street and still be a part of it. In contrast to Central Park which was envisioned by its nineteenth-century designers as a green oasis within which the surrounding dense urban fabric was often veiled, the character and layout of the High Line celebrates and embraces the experience of the city. A variety of designed platforms with seating direct and frame views onto the street life below and into the street canyons, evoking a sublime effect, similar to that described by turn-of-the-century writers when they wrote of New York City's streets framed by high rise buildings.

Much of the High Line's planting consists of species common in the city's hinterland. Imitating the spontaneous vegetation that had established itself between the deserted tracks, the plants provide a lofty ambience and a colorful picturesque frame for vistas between buildings and across roofs. Recalling the structure's former transportation use in the paving and furniture design, the High

Line aestheticizes the site's industrial past, but it does not hide it.

The lofty promenade is both set apart from and united to the city through these devices. Its design has turned a disused industrial railroad into an amenity. It both highlights and veils the former industrial character of the site and its development into a high-end tourist destination and a setting for the homes of wealthy residents. In this sense, then, the High Line can perhaps be considered a less duplicitous and more explicit landscape than the nineteenth-century design for *Central Park*. By recalling the site's history, the design for the adaptive reuse of the former freight line has set a precedent for the improvement of urban landscapes. •

满怀钦佩的回溯

马塞拉·伊顿，阿兰·塔特

　　我们初次知道吉姆[①]的著作是他 1990 年和 1991 年在《景观杂志》[②] 上的两篇关于当代景观理论的论述。1995 年 3 月，他在建筑协会举办的"景观的复兴"（确切名称不详）大会上发言。随后，吉姆又出版了两本学术著作——《测量美国景观》（1996 年）以及论文集《论当代景观建筑学的复兴》（1999 年）。这些成就促使我们邀请他在 1998 年秋天（那时对一个漂泊的曼彻斯特人来说还不算太冷）到曼尼托巴大学演讲，当时现场座无虚席。

　　那次周末的会面是令人感到美妙、有趣且尽兴的（葡萄酒都喝完了）——我们多次提到电影《冰血暴》——期间，吉姆强调他的理论研究只有通过景观作品才能得以验证。之后，他和他的 Field Operations 事务所宣布了在高线公园的实践资格，紧接着他们又设计了位于伦敦伊丽莎白女王奥林匹克公园的赛后广场、芝加哥海军码头和其他许多备受瞩目的项目，这些成

---

① 詹姆斯·科纳的昵称。——译者注
② 英文杂志：*Landscape Journal*。——译者注

就让人想起电影《偷天换日》中那句有名的台词——"你只能
杀出一条血路"。正如《伟大的城市公园》^①中有关高线的章节
所指出的，他的同事丽莎·斯韦金评论道，吉姆"致力于展现
景观的维度，提升风景园林的形象——风景园林是一门兼具理
论性和实践性的学科"。同时，丽莎一定也注意到吉姆在学界
和实践中作为导师的豁达品行。

　　根据我们对高线公园的观察，它展示了一种以风景园林
为导向的方法，将公园由外向内呈现出周围城市令人惊讶的全
景。我们的结论是："它例证了公园被视为城市及其变化过程
中不可分割的组成部分，而不是逃离城市的世外桃源。它证明
了为各种废弃工业用地寻找新未来的重要性，并表明巧妙的
完整性设计是对令人哀叹的、被遗弃的、真实性的一种有效
替代"。

　　2013 年，布莱恩特公园有限公司的执行董事丹尼尔·比
德曼指出，在高线公园项目取得成功后，吉姆和他的 Field
Operations 事务所将有资格参与纽约的每一个景观项目。伦敦
奥林匹克公园的客户代表菲尔·阿斯科夫则指出，吉姆很清楚
他在工作中想取得怎样的成就，而且下定决心要实现它。倘若
没有这样的决心，就没有今天的高线公园。

　　20 年前，吉姆和我们针对他探索理论工作和设计实践之
间共生关系的想法进行了交谈。如今，我们回顾他的成就（包

---

① 英文原著：Tate, A. *Great City Parks*. Routledge, 2015。——译者注

含设计项目和出版物），钦佩之情仍溢于言表。我们佩服他的幽默、慷慨和学识，在学界和实践中给予的指导以及他在阐释想法时清晰的图示语言，而这些影响了一代设计师。总之，他是一位才华横溢且令人钦佩的风景园林师。

Look Back in Admiration

Marcella Eaton and Alan Tate

Our first experience of Jim's work was his two discourses on theory in *Landscape Journal* in 1990 and 1991. Those papers preceded his presentation at the (questionably-named) *Recovery of the Landscape* conference at the Architectural Association in March 1995. This was followed by two of Jim's books that have become academic staples—*Taking Measures Across the American Landscape* (1996) and his collection of essays on landscape theory, *Recovering Landscape* (1999) . These publications prompted us to invite him to come and talk—to a full house—at the University of Manitoba in Fall 1998, before the winter became too bleak for a peripatetic Mancunian.

That visit was a wonderful, funny, wine-filled (and emptied) weekend—with multiple references to the movie *Fargo*—during which Jim stressed that his theoretical work could only be justified through built landscape works. And subsequently, in an achievement evocative of the infamous line from the movie *The Italian Job*— "you're only supposed to blow the bloody doors off" —he and his firm Field Operations announced their practical credentials with the High Line and then followed it with projects like the post-Games concourse at *Queen Elizabeth Olympic Park* in London, *Navy Pier* in Chicago and numerous other high-profile designs. As we noted in the chapter about the High Line in *Great City Parks*, his colleague Lisa Switkin has commented that Jim has "dedicated his career to demonstrating the scope, and elevating the profile of landscape architecture—as both a theoretical and a practical discipline". And Lisa would surely also note Jim's generosity as a mentor in academia and in practice.

Our own observation about the High Line is that it demonstrates a landscape architecture-driven approach that turns the park outsidein, presenting surprising panoramas of the surrounding city. And we concluded that "it exemplifies parks being seen as integral components of cities and their processes of change rather than as bucolic escapes from them. It demonstrates the importance of seeking new futures for all types of unwanted industrial site. It shows that design based on intelligent integrity is an effective alternative to lamenting abandoned authenticity."

Daniel Biederman, Executive Director of Bryant Park Corporation, noted in 2013 that after the success of the High Line, Jim and Field Operations would be justifiably considered eligible for every landscape project in New York. And Phil Askew, client representative for the *London Olympic Park*, noted that Jim knows exactly what he wants to achieve in his work and he is determined to achieve it. The High Line would not be what it is today without that kind of determination.

Now, nearly twenty years after our conversations with Jim about his intention to explore the symbiosis between his theoretical work and his design practice, we can all look back in admiration at his portfolio—at his projects and his publications. We can also admire Jim's sense of humour, his generosity of spirit, his scholarship, his mentorship in academia and practice, and the lucid graphic representation of his ideas that has influenced a generation of designers. An accomplished and admirable landscape architect. •

步道 – 天堂 – 巴比伦

阿德里安·高伊策

自 2009 年高线公园开放以来，它已成为当代风景园林和公园的国际典范。我们应该如何理解高线公园？她是一个公园还是一条林荫道？她对提升房地产价值有吸引力吗？

步道 – 天堂 – 巴比伦

乍看之下，高线公园是一条城市高架人行道，一条通道，一段探索天际线的旅程，它远离日常拥堵，将其从生活的痛苦中解放出来。在曼哈顿城市的基岩上，四季就这样循环交替着。这条步道不需要蜿蜒曲折，城市会为它让路。她被修剪得像个超模，精致、细腻，精力充沛。走在上面，你锃亮的鞋子不会沾上一点灰尘。人们在上面漫无目的地行走，自豪而坚定。

步道 – **天堂** – 巴比伦

在中世纪，天堂是经久不衰的艺术主题。在新的天堂叙事中，圣母玛利亚和她的孩子取代了亚当与夏娃的位置。上莱茵河画家的《天堂小花园》中，天堂就是主题。画中的花园被包围着，十分安全。圣母玛利亚在智慧之树的林荫下看书，她的身边有圣徒陪伴，他们相互关爱、吟诵、阅读、做园艺工作以及烹饪。在那里人们长生不老，四季如春，鸟语花香，浅草芬芳。同时，他们知道如何与猴子所代表的魔鬼共处。

毫无疑问，高线公园的花圃就如同这样一个天堂。那里有皮特·奥多夫留下的珍奇植物，有富裕慷慨的邻居为精细维护的种植坛买单，一种天堂的幻觉由此产生了。当然，在这个美国版的天堂里既没有蛇也没有猴子。高线公园孕育着一份纯真，她的风景优美动人，井然有序，张弛有度，规划合理。她向人们分享着和谐的世界，并保护自己免受奇怪的、突如其来行为的伤害。

步道 – 天堂 – 巴比伦

巴比伦是第三个隐喻，不是空中花园，而是巴比伦塔。创造和建造它的人害怕默默无闻和终有一死，他们热衷妄想，渴望出名和长生不老。巴比伦塔是无耻的虚荣宣言：这就是我们！他们的塔将直通天堂。它的建造者最终将达到上帝的高度。由于上帝的干预，从那时起，人们的语言不再相通，思想不再统一。

从寓意上讲，纽约向来是自由、资本主义、文化、欣喜、财富和权力的最佳代表。它的天际线高耸入云，它的街景挤满了无法相互理解的人。人们共享着城市空间，却互不相识。同样地，高线公园也想伸向天空，分享人们可以登上天堂的幻想。

著名的俄罗斯艺术家伊利亚和艾米莉亚·卡巴科夫创作了许多雕塑作品，其中就有一座巨大的天梯。人们爬上梯子，伸出双手，渴望遇见天使。在这些艺术作品中，他们讽刺地将巴比伦式的虚荣与希腊神话中的伊卡洛斯联系起来。

Walkway – Paradise – Babylon

Adriaan Geuze

Since its opening in 2009 the High Line has become the international reference for contemporary landscape architecture and parks. How should we understand the High Line? Is she a park or is she a boulevard? Is she an attraction for raising real estate values?

Walkway – Paradise – Babylon

On first glance the High Line is and elevated urban walkway, a passage, a journey which explores the skyline, excludes the daily congestion, and liberates itself from the pains of life. The seasons fly unchained from gravity high above Manhattan's bedrock. This walkway does not need to meander, the city vibrates. She is manicured like a super model, sophisticated, smooth, well drained. She will keep your polished shoes shiny. People walk proudly and firmly on their journey to nowhere.

Walkway – **Paradise** – Babylon

In the Middle Ages, the paradise theme was ever present in

art. In the new narrative of paradise Madonna and her Child took
the position of Adam and Eve. In *The little Garden of Paradise* by
the Upper Rhenish Master, paradise is the subject. The Garden is
enwalled and safe, the tree of wisdom drops shade for a reading
Madonna. She is accompanied by saints, characters which love,
chant, read, garden, cook. Nobody is ageing, springtime lasts
forever, trees and flowers blossom, herbs smell, songbirds sing.
They know how to live with the devil who is represented in the
monkey.

No doubt the High Line flowerbeds easily match this paradise.
Due to Piet Oudolf's unique botanic legacy and the unprecedented
generosity of rich neighbors who pay for these micromaintained
flowerbeds, people are given an illusion of paradise. Of course,
this American rewrite of paradise features snake nor monkey. The
High Line cultivates innocence. Her scenery is lovely, organized,
pre-programmed, and always reasonable. She shares the harmoni-
ous world and will protect against strange, non-scripted behavior.

Walkway – Paradise – **Babylon**

Babylon is the third metaphor, not the Hanging Gardens, but
the Tower of Babylon. Created and built by people who feared to
be anonymous and mortal, eager to pretend, to be famous and to
be immortal. The Babylonic Tower is the shameless statement of
vanity: this is us! Their tower would reach into heaven. Its build-

ers would finally equal God. Due to His interference, from then on people had to speak different languages, they no longer understand each other.

Allegorically, New York is definitely the best ever representation of this: freedom, capitalism, culture, euphoria, wealth, power. Its skyline peaks into heaven, its streetscape is full of people incapable of understanding each other. People share, but never meet. Equally, the High Line wants to reach into the sky, sharing the illusion people can climb into heaven. The well recognized Russian artists Ilya and Emilia Kabakov create sculptures, in which monumental ladders reach out into the skies. Human figures climb them and stretch out their hands, in desire of meeting an angel. In these artworks, they ironically link the Babylonic vanity with the Greek myth of Icarus. •

溯源

克里斯托夫·吉罗特

　　谁没听说过纽约的高线公园？在景观史上，从来没有这么少的树木却得到如此多的回报。在奇迹和幻想之间，高线公园雄伟地矗立在曼哈顿下城曾经的肉类加工区。几十年来，火车把牲畜和农产品从新泽西花园州运过来，然后在附近的仓库里屠宰和包装。这些古老的轨道能带我们去哪儿，这重要吗？如今，它已经成为城里最时髦的地方，吸引了无数的城市居民，他们每天穿梭于由曾经破败不堪的房屋改造而成的精品店、博物馆和咖啡馆。

　　高线公园是一个独树一帜的项目，因其形状、形式和功能而具有独特的地位。作为一个工业废墟，这座翻新过的铁路桥横跨曼哈顿下城的街道，它的铸铁支架与世界上任何其他的空中花园都不一样；它的作用更像是一艘船的甲板，把我们带到某个迷人的地方。但究竟是什么让高线公园如此与众不同？

　　高线公园的初期灵感来自柏林的轨道三角区公园和美山南麓公园以及埃森关税同盟煤矿区的标志性生态与后工业景观。高线公园从德国早期的城市生态学探索中汲取精华，例如粗糙的道碴、生锈的铁轨、渗有杂酚油的铁轨枕木，以及灌木丛生的荆棘和弯曲的树木交织而成的难以逾越的植被<u>丛</u>。如今，这些特征在高线公园的 DNA 中仍依稀可辨，成群的白桦树证明了这一点。但从本质上讲，由于其纯粹的都市风格，高线公园绝不是生态的或粗野的，它已经蜕变为一种完全不同的东西，类似于自然界的波将金 ^①，变得既时尚又虚假。这是一个分享新的自然崇拜的非凡之地，一条具有仪式性的林荫道，设置在通往奇幻之地的华丽小径上，在那里，世俗的曼哈顿人在精致的花朵和城市家具中，努力寻求自己内心的救赎。高线公园首先是一个社交场所，在这里，重要的是热爱自然和被别人看到。很少有游客记得它曾经的样子，那时干燥的苔原植物覆盖着废弃的铁轨。这只是一个极端天气盛行的迹象，北极的风寒到来时，这里的气温可以降到零下 50 度，而夏季，城市闷热的温度高达 40 度。这并不是什么奇闻轶事，因为当我们谈及高线公园植物的根系时，尽管它们可能又少又浅，却永远暴露在恶劣的天气中，这是其他城市景观未曾有过的挑战。因此，詹姆斯·科纳和皮特·奥多夫年复一年地为高线公园带来了春天的希望，实在是一项巨大的工程。

---

① 常用来嘲弄那些看上去崇高堂皇实际上却空洞无物的事物。——译者注

Back to the Roots

Christophe Girot

Who has not heard of the High Line in New York? Never in the history of landscape architecture was so much owed to so few trees. Between miracle and mirage, the High Line looms defiantly over what used to be the Meat Packing District of Lower Manhattan. For decades, trains used to carry livestock and produce in from the Garden State, to be slaughtered and packed in the adjacent warehouses. Where do these old tracks take us anyway, and does it even matter? It has now become the hippest place in town, attracting countless crowds of city dwellers, seeking day in and day out the attraction of boutiques, museums and cafés that have sprung-out of the once decrepit tenements.

The High Line is a project that stands out on its own, with a unique stature because of its shape, form and function. As an industrial ruin, the refurbished railroad bridge straddling the streets of Lower Manhattan on its cast iron legs comes across differently than any other hanging garden in the world; it works more like the deck of a ship taking us to some place of enchantment. But what exactly makes the High Line so different?

Early inspiration for the High Line came from the iconic ecological, post-industrial landscapes of the *Gleisdreieck* and *Schöneberger Südgelände* in Berlin and the *Zeche Zollverein* in Essen. The High Line drew its substance from the ruderal traits of these earlier German experiments in urban ecology, such as cruddy ballast, rusted tracks, tagged rail ties oozing with creosote, and sprawling bramble with shrubs and crooked trees woven into an impenetrable entanglement of vegetation. These traits are still vaguely discernible in the DNA of the project today, the clusters of birches attest to that. But in essence the High Line is anything but ecological or ruderal, because of its sheer urbanity; it has metamorphosed into something quite other, akin to a Potemkin of nature, that has become as fashionable as it is fake. It is an extraordinary place to partake in a new cult of nature, a ritual promenade set on some flowery path to OZ, where the mundane Manhattan crowd, amid exquisite blooms and furniture, strives for its own immanent salvation. The High Line is first and foremost a placed to socialize, where it is important to be seen and to be well in love with nature. Few of the visitors probably remember how it once was, when dry tundra vegetation covered the abandoned tracks. It was just an indication of the extreme conditions that prevailed up there on deck, with arctic winds and cold down to minus fifty degrees alternating seasonally with sweltering urban heat up to plus forty. This is no anecdote, because even as we speak the roots of the High Line, as few and shallow as they may be, are permanently exposed to the elements like no other landscape in the City. It is therefore quite a feat that Jim Corner and Piet Oudolf have achieved together by bringing back a promise of spring to the High Line year after year. •

新的后工业设计方法

沃尔夫拉姆·霍夫

　　高线公园是一个非凡且成功的城市开放空间。种植设计将园艺之美与城市自然联系起来，高架人行道提供着融入曼哈顿城市肌理的绝佳视野，使公园成为主要旅游景点。纽约人认为它是具有独创性的公园，但对风景园林师来说，高线公园是后工业时代景观设计语言结尾处的感叹号。

　　针对后工业景观设计的探讨始于 1975 年理查德·哈格设计的西雅图煤气厂公园。它是第一次对工业遗迹进行保留，并有意识地将其纳入公园设计中。实际上，这一设想违背了当地居民的意愿。公众认为自 1956 年以来位于布朗角的废弃煤气厂是城市衰败的象征，人们因此赞成彻底拆除工厂所有剩余结构，并在那里建造一座维多利亚风格的公园。然而，哈格对气化塔的美学赞颂标志着后工业遗迹创造性融入风景园林的开端。

　　尽管不断发展的美国环境保护运动主张处理那些被污染的场地，然而 10 年后人们对后工业设计的探讨却在欧洲首次取得了进展。在文化与美学层面，当人们对后工业遗迹的理解逐渐加深时，对新的城市生态学的兴趣也日渐浓厚。德国杜伊斯堡北景观公园将旧钢铁厂改造为休闲公园，它标志着从清理废弃工厂到发展后工业景观的范式转变。值得注意的是，当地政府将保护与整合废弃炼钢炉作为 1989 年设计竞赛的先决条件，而当时距离工厂关闭仅 5 年时间。

20 世纪末，在风景园林中利用废弃工业基础设施已是司空见惯。巴黎林荫道（1993 年）和柏林南麓自然公园（2000 年）成功地利用废弃铁路作为线性公园；现存自发生长的植被是种植设计的骨架，而高线公园的设计秉承了这一传统。事实上，高线公园之所以成为可能，是因为长达 30 年的后工业景观设计改变了公众对废弃工业基础设施的看法。20 世纪 70 年代中期，西雅图的居民希望拆除煤气厂；21 世纪，纽约人却爱上了位于肉类加工区的废弃高架铁路。乔尔·斯特菲尔德的摄影作品证实了人们将破败的基础设施视为与世隔绝的城市自然的看法。当这些高架桥计划被拆除时，一群有影响力的居民成功地游说，将其保存了下来。

高线公园标志着为适应性再利用这一传统掀开了新篇章。那些可见的废弃基础设施没有被保留，而是进行了彻底地重建。在进行公园设计时，人们发现高架铁路的结构存在问题，有必要彻底清除轨道和植被。然而，为了维持后工业的吸引力，轨道被重新铺设在新的枕木上。皮特·奥多夫的种植设计受到原先野生景观的启发，但却发展为一个维护良好且精致的园艺作品。因此，高线公园是超越后工业的，它为废弃基础设施提供了新的后工业设计方法。

A Neo-Post-Industrial Approach

Wolfram Höfer

The High Line is an extraordinary and successful urban open space. The planting design links horticultural beauty with urban nature, the elevated walkway offers amazing views into Manhattan's urban fabric, making the park a prime tourist attraction. New Yorkers think it is the first of its kind, but for landscape architects the High Line is the final exclamation point on the post-industrial landscape design discourse.

This discourse was opened in 1975 by Richard Haag's design for *Gas Works Park* in Seattle. This was the very first time that industrial relics were preserved and consciously integrated into a design for a park, in fact, against the wishes of local residents. The public considered the gas works at Brown's Point, abandoned since 1956, as a symbol of urban blight, and was in favor of completely demolishing all remaining structures to make way for a Victorian-style park. Haag's aesthetic celebration of the gasification towers marks the beginning of the creative integration of post-industrial relics into landscape architecture.

Although the evolving American environmental movement advocated for addressing contaminated sites, the post-industrial design discourse first gained momentum in Europe a decade later. A growing interest in novel urban ecologies went hand-in-hand with the evolving cultural and aesthetic appreciation of post-industrial relics. The paradigm shift from just cleaning abandoned factories to embracing post-industrial landscapes is marked by the *Landscape Park Duisburg-Nord* in Germany that transformed an old steel plant into a recreational park. It is noteworthy that local authorities made the preservation and integration of the abandoned steel furnace a prerequisite for the design competition held in 1989, only five years after the plant was closed.

By the end of the twentieth century, utilizing abandoned industrial infrastructure was commonplace in landscape architecture. The *Promenade Plantée* in Paris (1993) and the *Südgelände* in Berlin (2000) successfully utilized abandoned railroads as linear parks; existing voluntary growth was the backbone of the planting design. The High Line is designed in this tradition. In fact, the High Line was only possible because three decades of post-industrial landscape architecture had changed public perception of derelict industrial infrastructure. While Seattle residents of the mid-seventies wanted a gas works demolished, New Yorkers of the new millennium were in love with the abandoned elevated railroad in the Meat Packing District. The photographs by Joel Sternfeld supported the perception of crumbling infrastructure as secluded urban nature. When the viaducts were scheduled for demolition, a group of influential citizens successfully lobbied for their preservation.

The High Line marks a new chapter in the tradition of adaptive re-use. The visibly abandoned infrastructure was not preserved, but literally reconstructed. Developing the park design, structural issues of the viaduct were discovered that made a complete removal of tracks and vegetation necessary. However, in order to maintain the post-industrial appeal, the tracks were placed back on new railroad ties. The planting design by Piet Oudolf is inspired by the previous selfseeding landscape, but evolved into a very well maintained example of sophisticated horticulture. With that, the High Line is beyond post-industrial, it opens the door to a new neo-post-industrial approach to abandoned infrastructure. •

场所精神

约翰·迪克森·亨特

　　尽管有人呼吁在美国各地的废弃基础设施上模仿高线公园，但这无论如何是不可能的。高线公园是如此的"纽约"，甚至比它的设计师所预见的更好、更令人满意——这是一个有趣的例子，人们对它的接受程度超过了设计师的预想。高线公园证实了景观设计的基本法则，即对场所的尊重。虽然一些法国现代理性主义者认为，场所精神是不存在的，但是塑造一个完全自成一体且无法被效仿的地方，则是对场所这一古老概念的全新"诠释"。

　　事实证明，高线公园非常受欢迎。对于一些想利用其邻里优势的人来说，它可能太受欢迎了。然而，它的成功不仅仅因为这里是"纽约"，还因为它对废弃铁路所进行的富有想象力的转化是一种新概念——虽然有人怀疑巴黎的绿色长廊是它的前身，尤其是作为 19 世纪阿尔方 ① 的巴黎林荫大道的现代衍生品，但它们却十分不同。

---

① 阿尔方将英国自然风景园与法国规则式园林相结合，创造出风格杂糅的混合式城市公园。——译者注

　　高线公园通过对场地现状的赞颂和愿景改变了一处废弃基础设施。那些现实中的遗迹和具体形式代表着高线公园对这条旧铁路线的认可；它利用皮特·奥多夫设计的一系列奇妙的植物改变了废弃地上野草丛生的状况，却没有模仿植物过度生长的面貌，这些都让詹姆斯·科纳回忆起20世纪50年代比切姆切断铁路系统给英国所造成的破坏。事实上，公园改善了曼哈顿的肉类加工区，并给该地区带来了迅速的、实质性的改善。

　　高线公园吸引了很多游客，至少我每次去都是如此，这一点成为景观设计的一个关键性因素。公园为游客提供了一个城市剧场。花园和剧场之间的联系是长期存在的，它们既是表演的场所，也被视作表演者的场所，尤其是对于那些与观众互动的戏剧。高线公园至少从三个方面做到了这一点。

　　当然，还有可以俯瞰街道的露天剧场——这种结构经常被嵌入公园和花园中，但在这里，坐在高处的人们有机会凝视桥下穿梭的行人。

　　同样令人着迷且偶尔触动人心的是游客看与被看的方式，在任何一个时刻都不能确定你是看的那个还是被看的。甚至是那些不时出现在路线两旁建筑中的住户，他们既是熙熙攘攘的人流和那些纽约公寓的看客，也发现自己已成为公寓中那些"窥视者"的对象。

　　最后，高线公园带着人们穿梭于纽约，忽而转向，哈德逊河的景色尽收眼底：至此，对于未曾住在那里或是曾经栖居却已生活在别处的人来说，纽约本身就成为他们欣赏一出"好戏"的地方。

Genius Loci et Sui Generis

John Dixon Hunt

Despite the calls to imitate the High Line across the United States on other derelict infrastructures, it was for the best of reasons implausible. It was so satisfactorily "New York", better and more acute than perhaps even its designers foresaw - an interesting instance of reception outpacing a designer's vision. But it also acknowledged a fundamental law of landscape design that it honour place. Some modern French rationalists have argued that *genius loci* does not exist; but to make a site that is completely of itself, in the best of ways unimitable (*sui generis*), is just a new "take" on that ancient concept.

The High Line has proved enormously popular, maybe too much so for some who want to use its neighbourly advantages. Yet it is not just that it is "New York", for such an imaginative transformation of a derelict rail line was a new concept - some suspected that the *Promenade Plantée* in Paris was a predecessor; but that earlier example was quite different, not least for a being a modern spin-off from Alphand's 19th-century *Promenades de Paris*.

The High line transformed a derelict infrastructure by celebrating its local condition and aspirations: it acknowledged a former railway line with both actual remains and concrete representations of them; it transformed the weeds and wild growths on derelict sites – something that Jim Corner would recall form the devastations wrought in the UK by Beecham's cuts to the railway system in the 1950s – with a wonderful array of plants by Piet Oudolf, that recalled without mimicking overgrowth; it enhanced a section of Manhattan, the Meatpacking District, and brought quick and substantial improvements to the area.

And it also pulled in crowds, or did every time I visited it. That was also a key element of any landscape design that it provides its visitors with an urban theatre. The link between gardens and theatres is longstanding, as both a place to perform and be seen as performers, especially for dramas that spoke to their audiences. The High Line did that in at least three ways.

There was, of course, the amphitheatre of seats that overlooked the streets – a structure frequently inserted into parks and gardens; but here the chance was for those seated aloft to gaze down on the pedestrians below.

Equally fascinating, and occasionally a touch provoking, were the ways that visitors were always both looking and being seen, with no certainty as to which anyone was at any given moment. Even the occupants of the buildings that occasionally lined its route were spectators of the slow moving crowds, as well as the objects of *voyeurs* who found they could look into those New York apartments.

And finally, the High Line that took one through a segment of New York eventually turned towards a view of the Hudson: so New York itself became a sufficient play or drama for those who ether did not live there or who did inhabit, but lived elsewhere. •

新的城市自然

安德烈亚斯·基帕尔

提到詹姆斯·科纳就必然要谈及景观设计或他所定义的"景观都市主义"在当今所扮演的重要角色。从社会和存在的角度，詹姆斯·科纳以广阔的视野思考景观：它是我们赋予城市形态的方式，而最终景观是我们对于场所的识别，但它也涉及"我们与自然力量相联系"的整体存在状态。

詹姆斯·科纳和我都有着共同的双重经历：在深受工业化影响且高度城市化的地区长大，对自然有着深切的渴望。随着景观越来越成为一种"容器"，它承载着诸如社交和接触新的城市自然这样被忽视的需求与渴望，对景观的需求就是在这样的框架下被设定的。风景园林师应展开基于场所和人的策略，通过影响人们日常生活和地方特征的精准设计行动来接受这些挑战。

高线公园项目以一种全新的方式完美地体现了这些挑战，将人们聚集在一起，提高他们对城市更新这一文化转变的认识。由于市民的参与、城市规划和名人的支持，高线公园得以保存和改造。在这里，设计成为一种激发想象力的手段。用詹姆斯·科纳的话来说，"高线公园为公共生活中的欢乐、愉悦和戏剧表演提供了一个舞台。我们凭借曼哈顿的风光创造了一段旅程"。漫步在 30 英尺高的街道上方，穿过桥梁，发现曼哈顿上空迷人的景色并产生超现实的洞察力，它意味着能够通过改变视角来感知这一场所的身份性，从而获得一种全新的认识。2012 年，当我第一次参观高线公园时，我就能深刻地理解詹姆斯·希尔曼所说的："探索后现代城市需要利用新的感知形式，而不仅仅是设计形式"。这段经历让我们回想起伊塔洛·卡尔维诺具有启发性的见解，景观被视为人们讲述的有关场所的一系列故事。"不仅仅是一种艺术状态，景观似乎成为一种叙事，在过去、现在和未来中显现人与自然之间的复杂关系"。城市自然表达了其自身对逃避和连续性的需求，它在工业化进程中被遏制，现在正逐渐地重获新生并进入大众视野。

1859 年，在前工业时代，中央公园的建造主要是出于对社会福利的关注，并以市场为导向希望提高周围环境的经济价值。2009 年，在后工业时代，基于曾促使中央公园创建的相同社会经济动因，受到巴黎林荫道启发的高线公园首次向公众开放。这两个案例代表了以场所营造为主的城市变化的最佳范例，给全世界留下了深刻的印象。基于詹姆斯·科纳的经验，我和我的 LAND 事务所团队每天都在努力地重视重新连接人与自然，为当下和未来社会创造宜居场所。

New Urban Nature

Andreas Kipar

Talking about James Corner means inevitably talking about the crucial role that landscape design or, as he defines it, "landscape urbanism" is playing nowadays. James Corner's broad vision considers landscape both in social and existential terms: landscape is how we give form to our cities, ultimately how we identify with places, but it also deals with our whole existential condition of "beings that are bound with natural forces".

James Corner and myself share a common double experience: having grown up in a highly urbanised area severely impacted by industrialization, while feeling a deep need of nature. The need for landscape is to be set in this framework, as landscape is increasingly the recipient for unheeded needs and desires, such as socializing, and getting in touch with a new urban nature. Landscape architects should take on such challenges by deploying place-based as well as peoplebased policies through punctual design actions impacting people's everyday life and local identity.

The High Line project perfectly embeds those challenges in

a completely new way, bringing people together and raising their awareness of the cultural shift of urban regeneration. The High Line was saved and transformed thanks to a combination of citizens' engagement, city planning and celebrity support. Design here became a means to provoke the imagination. In James Corner's words "the High Line provides a stage for the joys, pleasures and dramas of public life. We've borrowed the landscape of Manhattan and created a journey". Walking thirty feet above the street, traversing bridges, and discovering amazing views and surreal insights above Manhattan means being able to perceive the identity of a place just by changing perspective and therefore gaining a completely new perception. When I first visited the High Line in 2012 I was able to deeply understand what James Hillman meant when he declared: "Postmodern cities need to be discovered using new forms of perception and not only of design". This experience brings us back to Italo Calvino's inspiring vision of a landscape seen as a series of stories that people tell about a place. "More than just a state of art, landscape seems to become a narrative in which the complex relationship between human beings and nature emerges in its dimensions of past, present and future". Urban nature expresses its own need for evasion and continuity, having been suffocated during industrialization and now regaining, little by little, its presence and visibility.

Central Park was built in 1859, in the pre-industrial era, mainly out of concerns of social welfare and as a market-driven prospect to increase the economic value of its surroundings. In 2009, in the postindustrial era, the High Line first opened to the public, inspired by the *Promenade Planteé* in Paris and based on the same social and economic reasons that once motivated Central Park's creation. Both cases represent superlative examples of place-making city transformations, that have had made a deep impression worldwide. What I and my team at LAND are striving for every day is based on James Corner's lesson: the importance of reconnecting people with nature to create liveable places for the societies of today and tomorrow. •

思想角逐

彼得·拉茨

　　现代主义建筑史的书籍里都有一张不同寻常的配图，它呈现的究竟是一个现代主义风格的作品，还是一栋建筑，一个瀑布？或者，它不过就是一张普普通通的照片？但是，图中散发着令人痴迷的力量，也象征着人类技术凌驾于自然之上。

　　最终，对生态的良知战胜了所有的迷恋，如今绝对不会容许一栋建筑物直接建造在密林深处的瀑布之上。詹姆斯·科纳在一场言辞犀利的演讲中展示了这张弗兰克·劳埃德·赖特流水别墅的照片。那场演讲是 AA 伦敦建筑学院的系列讲座之一，我也是在那里与这位来自宾夕法尼亚大学的英国人相识了。

　　此后，我见证了他担任宾夕法尼亚大学风景园林系主任一职，他在实践方面的出色工作让该校的这一知名专业更闻名遐迩。在我看来，他和他的 Field Operations 事务所也正是在那时取得了第一个巨大的成功——清泉垃圾填埋场改建项目。他成为风景园林设计的重要代表人物之一，树立了城市设计的先锋理念。詹姆斯·科纳从根本上为"景观都市主义"打下了深深的烙印。他强调："风景园林师必须在城市规划和政策中承担起责任！这一点在以往的工作领域中要求得过于有限"。

　　之后，高线项目出现了。一条横跨曼哈顿西区的高架铁路，十分紧密又充满变化地连接着 34 号大道和甘塞沃特街。高线改建是一个造价高昂的企划，也许因为当时特殊的设计要求尚无标准可循。该项目是幸运的，"高线之友"协会提供了资金上的保障，协会由富有社会责任感且财力雄厚的市民组成。他们参考了国际上工业遗迹改造的范例，尤其是从德国和法国的经验中汲取灵感，成功地阻止了高线铁路原先的拆除计划。

　　Field Operation 事务所的设计方案，以生锈铁轨为线索展开的形式语言，在国际竞赛中脱颖而出。事务所的"铁锈模型"，既体现了历史的延续又满足了实际的使用，这一设计亮点曾轰动一时。高线公园成为举世闻名的城市地标之一，也是 21 世纪风景园林行业的金字招牌之一。值得注意的不只是高线公园出色的设计本身，它也激发了许多艺术家和建筑师的创作灵感。无论是在风景园林和城市规划领域内，还是社会对待城市历史的态度，高线公园都标志着一个本质上的范式转移。

　　这个项目证明了，一段废弃的工业基础设施也可以为城市留下积极的印迹。与浪漫主义和城市现代化的刻板论调不同，高线公园带来的持续影响力锁住了资本与文化，让城市历史更加厚重。

　　从弗兰克·劳埃德·赖特到高线公园的出现——这是一场思想上的角逐。但是，我似乎在某些细节上依旧嗅到了流水别墅的气息……？

Ein geistiger Sprint

Peter Latz

Es gibt ein verrücktes Bild in den Publikationen zur Geschichte der Moderne. Zeigt es wirklich ein Objekt der Moderne, zeigt es ein Gebäude, einen Wasserfall oder ist es tatsächlich nur ein Bild? Eine begeisternde Kraft strahlt dieses Bild aus, signalisiert aber auch die Dominanz menschlicher Technik über die Natur.

Trotz aller Begeisterung regt sich das ökologische Gewissen, denn heute dürfte ein Bauwerk mitten im Wald, direkt über einem Wasserfall, gar nicht existieren. In einem eloquenten Beitrag stellte James Corner *Falling Water* von Frank Lloyd Wright in einer Vortragsreihe an der Architectural Association School of Architecture in London vor. So lernte ich diesen Engländer von der University of Pennsylvania kennen.

Später erlebte ich James Corner als Chair des Department of Landscape Architecture, wo er den renommierten Studiengang mit spannenden Projekten vorwärts brachte. Damals hatte er auch die ersten großen Erfolge mit seinem Büro Field Operations, darunter – für mich bedeutend – die Gestaltung der Mülllandschaft der

Fresh Kills. Er wurde zu einem der wichtigsten Sprecher für Land-schaftsarchitektur, als einer führenden Disziplin für die Gestaltung unserer Städte. James Corner prägte den Begriff des „Landscape Urbanism" wesentlich mit und formulierte vehement, was im Berufsfeld nur zurückhaltend gefordert wurde: Landschaftsar-chitektur muss Verantwortung in Stadtplanung und Politik überne-hmen!

Dann kam die High Line, eine Hochbahn, eingeschnitten in die Blocks des westlichen Manhattan, die mit großer Intensität und Varianz die 34. Straße mit der Gansevoort Street verbindet. Es war ein teures Vorhaben, vielleicht weil es für solche Anforderungen noch keine Standards gab. Das Projekt hatte Glück, war finanziell durch die „Friends of the High Line" abgesichert. Diese Gruppe engagierter, vermögender Bürger setzte sich, inspiriert durch die internationalen Erfahrungen im Umgang mit Industrieruinen, ins-besondere in Deutschland und Frankreich, erfolgreich gegen den geplanten Abriss der alten Hochbahn zur Wehr.

In einem internationalen Bewerbungsverfahren überzeugte Field Operations mit einem konsequent aus der Formensprache dieses rostigen Stücks aufgestellter Güterbahn entwickelten De-sign. Es war eine Sensation, dass Field Operations den Rost für ein Modell des Erhalts und der Nutzung in die Hände nahm. Es entstand eines der weltweit bedeutendsten städtebaulichen Merkzeichen und Aushängeschilder für Landschaftsarchitektur des

21. Jahrhunderts. Dabei sollte das Augenmerk nicht nur auf dem einzigartigen Design liegen, das auch Künstler und Architekten inspiriert. Die High Line steht für einen fundamentalen Paradigmenwechsel in Landschaftsarchitektur und Stadtplanung, wie auch im gesellschaftlichen Umgang mit Stadtgeschichte.

Sie beweist zudem, dass ein überkommenes Stück industrieller Infrastruktur hoch geeignet sein kann, einem Stadtviertel einen positiven Stempel aufzudrücken. Die nachhaltige Strahlkraft der High Line lockt Kapital und Kultur und macht die Stadt reicher an Erlebnissenentgegen romantisierenden und modernistischen Stadtklischees.

Von Frank Lloyd Wright zu diesem Ansatz der High Line – das war ein geistiger Sprint. Aber – sehe ich in einigen Details vielleicht auch noch den Geist von *Falling Water* ... ?? •

街道与天空之间

戴维·莱瑟巴罗

　　城市总是比我们理解的更加丰富多彩。诗人和作家将其描述为神秘的、野蛮的和梦幻般的，这要归功于这样一个事实，即城市总是蕴藏着比它最好的表现形式更多的内容。这一点不仅适用于都柏林、柏林、威尼斯或耶路撒冷这样的老城，也适用于里约、圣地亚哥和上海那些 20 世纪左右已形成自我风格的城市，纽约也不例外。当然，每个城市都有自己的特点、风格或情绪，这些都能在生活方式、持久的形式、建筑、花园和街道上体现出来。那些卓越设计项目的伟大之处——高线公园肯定是其中之一——不是取决于它们是否与城市氛围相契合，也不是取决于对紧迫问题的处理，而是取决于它们对当代生活中不可预见性的揭示。伟大的项目不仅证实了这一点，而且增强了现实性，仿佛通过某种奇特的创造性考古的方式，它们能够帮助这座城市重返往昔，变得更为完整。

　　高线公园之前，谁能想到纽约市的一条废弃基础设施带有朝一日会变成公共生活的新中心？项目的确完成得出色，但其视觉质量与人们对特定场所生活质量的关注相互关联。当然，它也是有价值的，其适宜性在所期待的景观意象中是显而易见的。当然，高线公园是一种新的城市形态，而其新颖性源于对历史条件的深刻理解。

正像它所改造的原有高架桥那样，高线公园一路穿过城市的历史结构，时而沿着以网格著称的纹理、时而与之相交。虽然项目富有成效，但它对现有规划模式的突破并不像典型的城市路段那样显著，因为都市性的一个新维度已被发现：街道和天空之间的一条线或一个层次，它更像是第二层，几乎和第一层（街道）一样具有公共性，但更开放。正如柏拉图以来许多作家所观察到的，如果把城市比作人体，那么高线公园既不是纽约的脚，也不是头；而是它的躯干，也许是胃、胸或是心脏。

可以把高线公园想象成一条蜿蜒穿过湿地的木栈道，只是略高一些。它的下面不是一个水世界，而是铺装和种植穿过一个容纳着巨大差异性却始终规整的城市网格，一个接着另一个，每个都有独立的所有者，具有各自利益和表达。在高线公园上方，一条由不同高度和轮廓构成的天际线，让清澈的蓝色天空具有了截然不同的品质，展现着统一性。

排除街道和天空，高线公园的中层地平线在这两者之间形成了一种更为厚实的景观，它不如第一层丰富多彩，也没有第二层那么统一①，但依旧合情合理，尤其是在道路转弯或路段发生变化时（例如在露天剧场），那里会出现另一个纽约，一个新的城市：拥挤却连贯，紧凑但宜居，人类的进取心和所做的荒唐事将变得微不足道，被自然的力量与节奏所支配。

---

① 第一层和第二层分别指街道和天空。——译者注

Between Street and Sky

David Leatherbarrow

Cities are always richer than our ideas of them. Poets and writers have described them as mysterious, savage, and dreamlike, thanks to the fact that they always harbor more content than even the best of representations can indicate. This is not only true for old cities, Dublin, Berlin, Venice, or Jerusalem; but also those that have come into their own in the last century or so, Rio, Santiago, and Shanghai. New York is no different. Of course each city has its own character, style, or mood, evident in ways of living as well as durable forms, buildings, gardens, and streets. The greatness of great design projects – surely the High Line is one – is determined not by their suitability to the town's atmosphere, nor their handling of pressing problems, but in their disclosure of unforeseen possibilities for contemporary life. Great projects do not merely affirm, they augment reality, as if by some strange manner of inventive archaeology they help the city become more fully what it had always been.

Before the High Line, who would have thought an abandoned

strip of infrastructure in New York City could become a new center for public life? Of course the outcome is beautiful, but its visual quality is tied to concerns for the quality of life in that particular place. Of course it is also useful, but its suitability is apparent in imagery that challenges expectations. And of course it is a new urban form, but its novelty arises out of a deep understanding of historical conditions.

Like the original viaduct it transformed, the High Line represents a single-minded cut through the city's historical fabric, running sometimes with, sometimes against the grain of the famous grid. While productive, its breaks from the existing plan pattern are less significant than those from the typical city section, for a new dimension of urbanity has been discovered: a line or level between street and sky, nearly as public as the first, but much more open, rather like the second. If the city can be compared to the human body, as so many writers since Plato have observed, the High Line animates neither New York's feet nor its head; instead, its mid-section, its stomach perhaps, possibly its chest, or heart.

Think of the High Line like a boardwalk snaking through a wetland, just a little higher. Instead of a watery world below, its pavings and plantings pass over a grid whose insistent regularity tolerates excessive difference, one property then the next, each a separate owner, interest, and expression. The clear blue above has the opposite quality, unity, for there all the various heights and profiles contribute to a single skyline.

Marginalizing the street and sky, the High Line's mid-level horizon develops a thicker landscape between those two, less varied than the first, not as unified as the second, but intelligible nonetheless, and revealing, especially when the route bends or the section changes (as in the open-air theater), for then another New York, a new city appears: not only congested but coherent, intense yet inhabitable, human enterprise and folly governed by the powers and rhythms of the natural world. •

过于成功?

莉莉·利卡

　　詹姆斯·科纳、Diller Scofidio+Renfro 事务所与皮特·奥多夫一同向公众呈现了一个达到极致水准的项目——高线公园。一个风景园林作品，能在短时间内冲上"必去景点排行榜"并始终名列前茅，是十分罕见的。假如在 20 世纪初期，高线公园一定是贝德克尔①旅行指南上的三星推荐景点。今天，互联网上的旅游推荐指数就是一个项目成功的凭证。显然，高线公园取得了非凡的成绩。

　　高线公园得益于原有的高架结构。可以在高空穿越一个垂直生长的城市，视线不仅能捕捉到更高楼层的室内，还能在高楼耸立之间以全景视角领略峡谷般的街道风貌，这是多么令人震撼啊！在巴黎，也有一个项目具备类似的结构优势。它是位于 12 街区的勒内·杜蒙绿色长廊，菲利普·马蒂厄和雅克·韦尔杰在 1988 年的设计中发挥了这一空间优势，据说也是全球首次将废弃高架铁路转变成 4 公里长的线性公园。与高线公园相比，法国案例在设计和园艺方面的确稍显

---

① Baedeker 或 Baedeker-Reiseführer，德国著名且历史悠久的旅行指南。——译者注

逊色。勒内·杜蒙绿色长廊的设计中规中矩，有些拘泥于形式，甚至只在个别之处体现了高架的优势。而高线公园的每一处景观都是精心设计的，采用了新的设计语言来表现工业遗迹的氛围。非常完美的设计和施工，甚至是过于完美无瑕了。次生植被、油漆斑驳的栏杆和生锈的旧铁轨，这些粗糙感都不见了。取而代之的是一个精致的组合，由铺装、户外设施、植物配置以及定点铺设的装饰性铁轨共同构成：不允许有一丝一毫的意外。然而，"意外之美"正是这个废弃地曾经的魅力所在。此外，也是因为这种完美（和媒体的推波助澜）让高线的设计元素不断地被复制。人们试图复制同样的成功。如果能把"抄袭"客气地称之为"引用"，那么当下的设计正充斥着对高线设计元素的引用。然而，唯有少数能够达到相同的设计水准，更罕见的是拥有像高线公园一样极具吸引力的故事背景：一个嵌入城市网络的结构、一段支撑城市繁荣的铸铁架构、一道折射城市历史的璀璨光芒。正是场地本身所具备的条件成就了高线公园，同时也是这个项目叙事的核心。

　　新公园的"极致"程度还体现在另一方面：它是绿色绅士化的典型案例。对此，迈克尔·布隆伯格和城市用地主管部门应当负有主要责任，高线公园沿线已经成为地标式豪宅的黄金地段。此外，还有一份调查显示，原先有日常休闲空间需求的住户正在被慕名而来的人潮取代："与周围其他城区以及同类公园相比，高线公园的使用者在种族构成方面过于单一，缺乏多样性"。住在高线公园附近的知名平面设计师史蒂芬·萨格迈斯喜欢在公园里晨跑，也热爱那里各种植物变幻多姿的自然美。但是，他必须在早上 7 点离开，因为之后那里便摩肩接踵。就此而言，高线公园也为它的成功付出了代价。

赖克尔，亚历山大（2016）：高线公园与大众公共空间理想，城市地理，第 37 卷，第 6 期，2016，pp. 904–925.

金塔纳，玛丽埃拉（2016，8，8）：改变架构，探索高线公园的影响，http：//streeteasy.com/blog/changing–grid–high–line/（2018–01–14）.

Kann Landschaftsarchitektur *zu* erfolgreich sein?

Lilli Lička

Mit dem High Line Park legen James Corner, Diller Scofidio + Renfre und Piet Oudolf ein Projekt der Superlative vor. Selten schießt ein Werk der Landschaftsarchitektur auf der Hitliste der Must-See-Destinationen derart schnell und dauerhaft nach oben. Anfang des 20. Jahrhunderts hätte das drei Sterne im Baedeker bedeutet, heute ist die Zahl touristischer Online-Tipps der Beleg des ultimativen Erfolgs. Das ist eine außerordentliche Leistung.

Der High Line Park profitiert in besonderem Maße von dem alten Gefüge. Es ist aufregend, einen gewachsenen Stadtraum in Hochlage zu durchqueren, Ausblicke nicht nur in die Innenräume höhergelegener Etagen der Gebäude zu erheischen, sondern auch die Morphologie der Stadt mit phänomenalen Blicken in die Straßenschluchten zu erfassen. Eine solche Standortgunst haben Philippe Mathieux und Jacques Vergely 1988 für die *Promenade Plantée* im 12. Pariser Arrondissement ausgenutzt, indem sie – angeblich weltweit erstmals – die Trasse einer ehemaligen Hochbahn in einen 4 Kilometer langen Park verwandelt haben. Gestalterisch

und gärtnerisch steht das französische Beispiel dem amerikanis-
chen jedoch nach, es ist konventionell und teilweise formalistisch,
sogar die einzigartige Hochlage kommt nur an manchen Stellen
zur Geltung, während in New York Ausblicke aufwändig inszeniert werden. Die präzise Neugestaltung übersetzt die Atmosphäre
der Industrieruine in eine neue Sprache. Die gestalterische und
handwerkliche Umsetzung ist perfekt, vielleicht zu perfekt. Die
Rauigkeit der Ruderalvegetation, der vom Geländer abblätternde
Lack, die angerosteten Gleise sind ersetzt durch eine exakte Kom-
position von Belag, Möblierung und Bepflanzung mit akkurat ge-
setzten Gleisstücken: nichts ist mehr dem Zufall überlassen, der
den Charme des verkommenen Objektes zuvor ausgemacht hatte.
Es ist sicher auch diese Perfektion (und die mediale Präsenz),
welche die Elemente kopierbar macht. Dem Erfolg will nachgee-
ifert werden. Zahlreiche Zitate poppen in aktuellen Gestaltungen
auf, wenn man die Plagiate denn so freundlich nennen darf. Selten
haben diese die gleiche gestalterische Präzision, noch viel seltener
haben sie den attraktiven Kontext der High Line: eine ins Stadt-
gefüge eingewachsene Struktur, die Gusseisenkonstruktion einer
arbeitenden Stadt, die Ausstrahlung eines Ortes mit Geschichte.
Auch diese materiellen Voraussetzungen tragen zum Erfolg bei,
zumal sie zentral im perpetuierten Narrativ des Projektes stehen.

Der neue Park auf der High Line spielt allerdings auch in einem anderen Stück der Superlative eine zentrale Rolle: Sie wurde

zum Paradebeispiel der grünen Gentrifizierung, an der die Umwidmung durch Michael Bloomberg und die Stadtverwaltung erheblichen Anteil hat. Der Standort wurde zur ersten Adresse, ikonische Wohngebäude brüsten sich mit ihr. Einer anderen Untersuchung zufolge werden ansässige Erholungsbedürftige vom Ansturm verdrängt: die ethnische Zusammensetzung der Besucherinnen und Besucher entspreche nicht jener der Umgebung und sei auch nicht so durchmischt wie in vergleichbaren Parks. Der bekannte Grafiker Stefan Sagmeister benutzt die High Line begeistert für seine Morgenläufe. Er schwärmt von der natürlichen Pracht der sich ständig ändernden Vegetation. Allerdings muss er um sieben Uhr los, um sich dort noch bewegen zu können. Auch in dieser Hinsicht sei die High Line Opfer ihres eigenen Erfolges. •

Reichl, Alexander J. (2016): The High Line and the ideal of democratic public space, in: *Urban Geography*, Volume 37, Issue 6, 2016, pp. 904–925.

Quintana, Mariela (2016, August 8): *Changing Grid: Exploring the Impact of the High Line*, http: //streeteasy.com/blog/changing-grid-high-line/ (2018-01-14).

绿色与希望 ①

弗兰克·罗尔伯格

　　伴随着时差的疲惫，漫步在纽约中央公园。天呐，中央公园是如此的郁郁葱葱，往来者络绎不绝。倦意逐渐褪去，我开始放松下来，沉醉于这里的一切。草地边上有个卖热狗的小餐车，我并不喜欢吃热狗，吸引我的是看似无边无际的草坪。

　　在归零地②感受到的则是焦虑不安。9/11撞击的电视画面浮现在脑海里，让我回想起30年前第一次看到的世贸大厦。曾让我感动并铭记的还有与众不同的华盛顿越战纪念碑。眼前的归零地遗址，巨大又华美。概念设计、施工和细节都精益求精。但真正令我震撼的，是深埋在9/11地下博物馆里陈旧的地基挡水墙③。

---

① 作者在本文中采用了德国浪漫主义文体。原文句子结构类似于诗歌，省略了大量的句子成分，为确保句意的表达，本文采用了意译的方式。全文结构以讲述回忆的形式，把纽约最具代性的绿地中央公园、"9/11"撞击后的归零地遗址和高线公园串联在了一起。——译者注
② 归零地，纽约"9/11"世贸中心遗址上建造的纪念广场，高线公园位于中央公园和归零地之间。——译者注
③ 混凝土浇筑的地基防水墙（Spundwand）。用于防止紧邻的哈德逊河河水渗入和侵蚀地基，确保原世贸大厦的稳固性。"9/11"事件之后，地基防水墙被保留，成为9/11纪念博物馆地下展出的重要部分。参见"9/11"纪念博物馆官网。——译者注

最后，来到了高线公园。先前的阴霾戛然而止。一个充满活力的城区，有着众多的咖啡馆和游客。纽约在这里自我更新，变得生机勃勃、放眼未来，回归自然的绿色。到访过高线公园的人，都这样评价它，我也不例外。

高线公园在哪里呢？必须去寻找，拾级而上。不过，通往旧高架铁路的楼梯并不多，就像是"通往绿色天堂的阶梯"。城市的绿色空间就应该这般精心设计，与随处浇筑的钢筋混凝土形成鲜明对比。请不要再种更多的玫瑰在交通绿化带上！请不要再没有灵魂地绿化城市边缘！请看一看这里：高线公园是一座花园。来到这里的每个人都是带着计划和想法的。走上楼梯，就像是跨入花园的大门，让此处变得意义非凡。我喜爱这里色彩缤纷的植物，尤其是6月里翠绿的叶片。浅色白桦树配上深色的蒿草，肾形草、萱草搭配山桃草，都是常见的植物配置，它们被养护得非常好。黄栌则显得与众不同，叶子稀稀疏疏，长得并不秀美，凌乱又孱弱，花序远远地挂在枝头上。偏偏就是这黄栌的花序，让人拍案叫绝。请在网上搜索一下"高线黄栌"。

黄栌的花序是铁锈玫瑰色，与镶嵌着无数铆钉的铁桥还有沿线旧厂房的红砖外墙遥相呼应。黄栌花序，犹如昔日列车驶过时飘出的蒸汽云朵。这也是设计师的妙笔吗？

另一处，坐在木制平台上的人们，正俯视着西边17号大街的车辆。许多黄色出租车，走走停停。时不时有汽车鸣笛声，穿透一人高的玻璃隔墙。他们欣赏着眼前的街道。这一幕这让我联想起，自己曾经因为在家乡的高速公路上建造观景点，而不得不备受批评。甚至，某个电视节目主持人指责我浪费税收。他真应该来这里坐一坐，瞧一瞧，就像旁边的人一样，望着川流不息的车辆。

真的就像 Field Operation 事务所网站上写的那样？高线公园，即城市自然是"促进投资"吗？这个想法确实很美好。在灰色城市里被绿色空间占领的地方，的确会重新让人感到幸福，想要拿起手中的工具和颜料去打造一个更好的世界。但是，有的人说："既然禁止在高线公园上面加建，它相应的建造量就应当作为开发权转嫁到旁边的地块上，允许建造更高的大楼"。事实上，那些人就是这样做的。听上去是多么的狡猾和算计。世界似乎本不该如此。

我盯着铺装许久，观察那些预制混凝土枕木是如何优雅地升起变成了座椅，又是如何朝着植物的方向逐渐变细；铺装之间的缝隙是怎样由窄变宽，再渐渐地嵌入绿色植被之中。突然眼前闪现出一个画面，金属线条从天空中滑落，在世贸大厦入口处的外墙上方逐渐变细，形成虚实相间的结构样式，如同我刚刚观察的脚下铺装。曾经的世贸大厦外墙，只剩下 9/11 博物馆门口展示的悲伤残垣①。我是不是在高线公园看到了在设计形式上对此的呼应呢？放在这里应该是合适的，因为可以引发人们无限的遐想，去思考城市脉搏中的生命、绿色与希望。

---

① 这里指代的是世贸中心双塔被炸毁前，标志性的建筑外墙钢架结构——三叉戟双异型柱。——译者注

Über die Hoffnung und das Grün

Frank Lohrberg

Mit Jetlag im *Central Park*. Mein Gott, der *Central Park*. So viel Grün, so viele Menschen. Müdigkeit weicht Leichtigkeit. Ich will eintauchen in das Ganze hier. Hot Dog auf einer Wiese. Ich mag keine Hot Dogs. Aber dieses Gras. Diese endlosen Rasen. Beklemmungen am *Ground Zero*. Fernsehbilder im Kopf. Gedanken an den ersten Besuch vor 30 Jahren. Und das *Vietnam-Memorial* in Washington, das mich anrührte, weil es so anders war. *Ground Zero* ist groß. Edel. Alles korrekt. Konzept, Ausführung, Details. Kein Fehler nirgends. Habe meine Gefühle im Griff. Bis zur alten Spundwand tief unten im 9/11 Museum.

Endlich zur High Line. Trübsinn bleibt zurück. Ein lebendiges Viertel. Viele Cafés, noch mehr Touristen. Hier erfindet sich New York neu. Lebendig, nach vorne schauend und natürlich grün. Liest man, sagt man. Sage ich auch.

Wo ist die High Line? Sie will entdeckt werden. Erklommen. Wenige Aufgänge hinauf zur alten Trasse. Stairways to a green Heaven. So funktioniert Grün in der Stadt. Nicht ausgegossen überall. Sondern gezielt gesetzt. Im Kontrast zu Beton und Stahl. Bitte keine Rosen mehr auf dem Mittelstreifen, keine seelenlose Eingrünung am Ortsrand! Schaut her: die High Line ist ein Garten. Wer kommt, kommt mit Absicht. Und Bedacht. Geht hinaufwie durch eine Gartenpforte. Das macht den Ort so wertvoll. Ich mag die Bepflanzung. Üppig. Vor allem Blattgrün, im Juni. Dunkle Seggen unter hellen Birken. Heuchera, Taglilien, Gaura. Routiniert komponiert, bestens gepflegt. Und dann der Perückenstrauch. Wow. Seine Blütenstände leuchten von viel zu weit oben. Kein schöner Wuchs. Zerzaust und stakelig. Zu wenig Blatt am Holz. Aber diese Blüten. Akrobatisch. Googeln Sie mal „Cotinus High Line".

Das Rost-Rosa der Blüten harmoniert mit den tausendfach genieteten Eisenbrücken. Und den Backsteinfassaden der alten Fabriken, die die High Line einst belieferte. Cotinus-Blüten als Dampfwolken einer vergangenen Zeit. Hat das die Planer bewegt?

Da sitzen Leute aufHolzpodesten und schauen hinunter aufdie 17th Street, West. Autos, Taxis. Viele gelbe Taxis. Stopp und Go. Ab und an dringt ein Hupen durch die mannshohen Glasscheiben. Die Leute schauen. Daheim musste ich mich schelten lassen für einen Aussichtspunkt, den wir an einer Autobahn bauten. Ein Fernsehmoderator warf mir Verschwendung von Steuergeldern vor. Er soll herkommen! Und sich hier hinsetzen und schauen, wie die Leute schauen. Auf Autos. Und Taxis.

Ist die High Line, ist Grün in der Stadt ein „catalyst for investment", wie es auf der Webseite von field operations heißt? Das wäre schön. Und bestimmt ist es auch so. Wo Grün die grauen Städte kapert, fühlen sich die Menschen wieder wohl, greifen sie zu Werkzeug und Farbe. Und bauen eine bessere Welt. Jemand sagt, die High Line darf nicht mehr überbaut werden. Das verlorene Bauvolumen geht als Baurecht auf die Nachbargrundstücke über. Dort könne man nun höher bauen. Was man auch macht. Das klingt schnöde und berechnend. So darf die Welt nicht sein.

Lange betrachte ich den Bodenbelag. Waschbetonschwellen. Elegant, wie sie sich zu Sitzen erheben. Und wie sie sich verjüngen zum Grün hin. Wie die schmalen zu breiten Fugen werden. Für das Grün! Mir schießt ein anderes Bild in den Kopf. Metallstreifen gleiten aus dem Himmel herab, verjüngen sich über dem Eingang und bilden das gleiche Muster aus Masse und Leere wie hier zu meinen Füßen. Von der Fassade des World Trade Centers blieb nur ein trauriger Rest im Entrée des 9/11 Museums. Sehe ich hier ein Zitat? Es würde zu diesem Ort passen, zumindest zu den ungezählten Gedanken, die er bei seinen Besuchern hervorruft. Über das Leben in einer pulsierenden Stadt. Über die Hoffnung und das Grün. •

蜿蜒曲折的纽约

瓦莱里奥·莫拉比托

除了对高线公园设计优点的个人见解外，这一项目最重要的是为我们提供了一种风景园林学科的当代创新性视角。作为世界上参观人数最多、最受欢迎和最上镜的地方之一，高线公园向每个人揭示了风景园林的能力，它可以创造更多民主之地，在当下和未来为我们的城市塑造身份性特征。

前段时间，詹姆斯·科纳赠送了我一本《高线公园》[①]，书中详尽讲述了整个项目的酝酿过程，并汇集了最初的草图、施工图以及项目使用的照片。赠书当天，我们重点讨论了风景园林学科的现在和未来、教学内容并思考谁有可能在未来城市的发展中起主要作用。我认为有三个词主导了我们的探讨：想象力、美学和精确性。风景园林与艺术的关系产生了创造力；诗歌、绘画、文学都是风景园林想象力的媒介。美，是结合经验和直觉、综合体验和冒险突破的想象形式。精确性，即万物皆有其确切的形式、技术、材料、生态性以及最终具体的表现，它是实现想象力与美学相互融合的必要因素。

[①] 英文原著：Corner, J. *The High Line*. Phaidon Press, 2015. ——译者注

　　当从纽约市的地面层进入位于 34 街和 11 街之间的高线公园时，便产生了一种特定的归属感。一个简单的、非纪念性的入口将游客带入一条倾斜、平缓和弯曲的混凝土道路，铁轨雕刻着路面，它们仿佛是被几代人遗忘的线条。从入口处，高线公园就开始了它的"摄入"和"吞咽"①，并将游客推向公园南端的更高层，它让你慢慢感受到自己是独特的城市"消化过程"的一部分。高线公园横断面的简洁性揭示了改变纽约城市普通认知的意图；它将游客引入一种不真实的体验，仿佛是一个增强的虚拟现实。穿过它，游客的感知被修改和改变，准备探索一个意想不到的地方，一个未知的"蜿蜒曲折之地"。人们边游览边拍照，防止遗忘，多样、复杂的空间沿着"羊肠小径"时隐时现。高线公园的特色预制混凝土铺装产生了动态的阴影，随着一天中不断变化的光线，这些阴影似乎描绘出精致的书法艺术。长椅、曲面、木桌和其他要素精确地丈量着空间；铁轨根据植被与游客的关系产生了复杂的几何图案组合。在游览过程中，触摸植物对人类起着至关重要的作用，它甚至可以抚慰心灵、追逐梦想并拓展想象力。伊塔洛·卡尔维诺②认为，他所拥有的想象力和创造力就像是一个以知识为"食物"、以文学为"成果"的消化过程。高线公园将我们"摄入"其中，改变着我们（也许是改善了我们），并让我们回归到纽约"正常"的生活之中。

---

① 将游客进入高线公园的游览过程比喻为一种食物进入人体的消化过程。——译者注
② 意大利当代作家。——译者注

Intestinum New York

Valerio Morabito

Beyond personal opinions about the merits of its design, the High Line most importantly has endowed us with a contemporary, innovative vision of the discipline of landscape architecture. Becoming one of the most visited, popular, and photographed places in the world, the High Line revealed to everybody the capacity of landscape architecture to create democratic places for the present and future identities of our cities.

Some time ago, James Corner gave me a copy of the book *The High Line*, which is an exhaustive telling of the entire gestation of the project, bringing together initial sketches, working drawings, and photographs of the project in use. The day he gave me it, our conversation was focused on the present and the future of the discipline of landscape architecture, what to teach and speculations on who might play a leading role in the future development of cities. Three words, in my opinion, dominated our discussion: imagination, aesthetics, and exactitude. Relationships with arts produce creativity; poetry, painting, literature are all agents for landscape architectural imagination. Aesthetics is the form of imagination that combines experience and intuition, consolidated experimentations and hazardous breakthroughs. The exactitude is the very same exactitude that as a result of which everything has its precise form, technique, material, ecology, and, ultimately, specific representation. Exactitude is the ingredient needed to achieve that alchemical combination of imagination and aesthetics.

There is a certain domestic feeling when entering the High Line from the ground level of New York City, between 34th and 11th Street. The simple and non-monumental entrance brings the visitors along an inclined, gentle, and curved concrete path, engraved by tracks like lines forgotten by lost generations. From this entrance the High Line starts its ingestion, swallowing and pushing visitors towards the higher level of the park to the south. It is a slow approach that makes you feel you are part of a unique process of urban digestion. The simplicity of the transect of the High Line reveals the intention to change the normal perception of New York City; it introduces visitors to an unreal experience, as if it were an augmented and virtual reality. Passing through it, the visitors' perception is modified and altered, ready to explore an unexpected place, an unknown land of an *intestinum*. Taking photos to protect the memory from being overloaded with impressions, a multitude of complicated spaces appears and disappears along the *intestinum*. The characteristic High Line precast concrete paving produces shades in forms of dynamic signs, painting sophisticated calligraphies in collaboration with the changing sunlight over the day. Benches, curved surfaces, wooden tables and other elements precisely measure the space; rails generate complex combinations of geometric patterns in according with vegetation, that establishes relationships with the visitors. Touching the vegetation plays an essential role in the digestion process, soothing souls, improving dreams, and expanding the imagination. Italo Calvino believed that his imagination and creativity were a process of digestion based on knowledge as food and literature as expulsion. The High Line ingests us, the incredible multitude, changing and returning us — perhaps improved — to the "normal" life of New York City. •

一种当代原型

乔·努内斯

　　因为高线公园项目与当代文化的利益密切相关，所以它对任何一个有关当代风景园林的调查研究来说都是极其重要的。它向大众介绍了项目的多重主题，尽管它们已经是景观文化的一个组成部分，但尚未超出风景园林的学科范畴；因此，高线公园成为一种传播景观思维方式和组织方式的焦点。

　　这些交叉的主题以一种简单且完整的方式清晰地构建了一个综合体，使高线公园项目在当下可被视为一种原型，一个极具表现力的教学案例，但作为一个概念化模型，它也同样有趣。

　　该项目涉及的主题有城市剩余空间的再利用、孔隙和结构的重要性，它们在过去往往被忽视；基础设施作为一种构建景观的工具；细节，作为一种设计通用语言，强调了自然对人类空间的自发占有。

　　该项目讨论了景观设计中的多个议题、景观设计的系统性条件，以及它们如何促进（外部／内部、开放／封闭、新／旧、基础设施／休闲）空间的融合，这些空间曾经很容易被一种将城市结构图解为更广泛类别的文化所区分。

A Contemporary Archetype

João Nunes

The High Line project became central to any survey of contemporary landscape architecture because it is deeply related to the interests of contemporary culture. It introduced thematic crossovers to a general public which, although they were already an integral part of the landscape culture, had not yet arrived outside of the discipline; the High Line became, therefore, the focus of the spreading of a way of thinking about and making landscape.

The crossover themes clearly constitute a synthesis in such a plain and complete way that the High Line project can now be considered as an archetype, extremely expressive as a didactic tool, but no less interesting as a conceptual model.

The project deals with the themes of re-use, interstice and the structural importance of spatial leftovers that often end up as secondary to other issues; infrastructure as an instrument for constructing landscape; detail, as a design universe celebrating the spontaneous appropriation of human structures by the organic.

The project talks about the overlapping layers in landscape

design; about the systemic condition of landscape design, and how it might promote the fusion of spaces (exterior/interior, open/ closed, new/old, infrastructural/leisure) once easy to distinguish in clear categories by a culture that schematized the city structure into larger groupings. •

相对紧凑却广泛相关

约格·雷基特克

　　在面对城市中的废弃铁路时，高线公园项目已成为世界上任何一个风景园林师都不能忽视的参照物。我带领两个工作室，与新加坡国立大学风景园林专业的学生一起设计了新加坡的旧马来亚铁道。它是一条总长 26 公里的狭长地带，从这一城市化岛屿的北部向南延伸。马来西亚和新加坡之间的土地交换协议早已生效，位于新加坡但属于马来西亚的土地已回归新加坡所有。这条铁路廊道有着茂密的热带植被，连接着成片的灌木丛、次生林、草原、红树林和滩涂，是一道迷人的城市景观。

　　在第一个设计中，学生们开发了一条"新加坡游径"，并在新加坡城市再开发部门发起的开放性创意竞赛中荣获一等奖。该设计体现了我们的坚定信念，一条旧铁路上的公园式游径对于土地稀缺的新加坡来说是非常有意义的。我们的第二个设计重新审视了该廊道的发展愿景，将其命名为"国家步行道"，其中包括将旧铁路转变成一条贯穿全国的无障碍超级连接通道；一条 200 万人使用的、对生态有益的无机动车通勤与社交地带。我们曾有着很大的梦想，如今却被一笔划破。一个前所未有的机会是可想而知的，但政府却没有发现这个获得全球认可的历史机遇。他们把连续的廊道切碎，其中大部分作为建筑再利用，并对剩余的几个破碎地块敷衍地发起了一次总体规划竞赛。在纽约，不仅仅是权威人士或设计师，"高线之友"协会也挽救了高线公园的残存结构，使之免于拆除，这一由约书亚·戴维和罗伯特·哈蒙德发起的基层运动，部分得益于乔尔·斯特菲尔德极具魅力的照片，它们在一年时间里记录了这个注定要被拆除的高架结构。在詹姆斯·科纳的指导下，高线公园项目一经开放就几乎立刻变得著名且广受喜爱，这座城市创造了当代城市景观设计中最令人难忘的范例。

　　当我问我的学生这个项目的长度时，他们并不确定，但有人估计它长达 15 公里。事实上，这段重新设计的高线公园只有 2 公里多。考虑到这些重要的城市基础设施在全球的突出地位，我们不得不承认，一个设计优秀、执行良好且空间相对紧凑的项目可以实现巨大的相关性。高线公园的故事是伟大的，因为它证明了城市公共空间质量可以成功地由它的朋友和使用者来决定。如果没有城市居民提供的重要灵感，管理者和设计师可能已经失败了。

Comparatively Condensed yet Colossally Relevant

Jörg Rekittke

The High Line project has become a reference that no landscape architect in the world can ignore when faced with an abandoned railway in an urban context. With my landscape architecture students at the National University of Singapore, I led two studios dealing with the former Keretapi Tanah Melayu (KTM) railway in Singapore, a strip of land twenty-six kilometers long, stretching from the north to the south of the urbanized island. A land-swap agreement between Malaysia and Singapore had come into effect, meaning the Singapore-located but Malaysia-owned land returned to the fold of the Singaporean nation. Host to lush tropical vegetation, the railway corridor connected patches of scrubland, secondary forest, herbaceous grassland, mangroves and mudflats: an urban landscape of fascinating beauty.

In the first studio, the students developed *The Singapore Trail*, later awarded first prize in an open ideas competition launched by the Urban Redevelopment Authority of Singapore. The design was a manifestation of our adamant belief that a park-style trail on the former railway would make perfect sense in land-scarce Singapore. Our second studio looked afresh at the development of a vision for the corridor, which we entitled *The National Mall*. This involved the transformation of the former railway corridor into a barrier-free super connector running the length of the country; a car-free tract for ecologically beneficial commuting and the social interaction of two million people. We had big dreams, but they were burst by the stroke of a pen. An unprecedented opportunity was palpable, but the government did not see this historic chance for global recognition. They cut the continuous corridor into pieces, repurposing the majority of it for building, and half-heartedly launched a master plan competition for the few scraps that remained. In New York City, it wasn't an authority or designer that saved what remained of the High Line structure from demolition, but the "Friends of the High Line" grassroots movement, initiated by Joshua David and Robert Hammond, who were helped in part by Joel Sternfeld's charismatic photographs that documented the doomed structure over the course of a year. After it opened the High Line project almost instantly became famous and well-loved—guided by James Corner, the city had created what turned out to be the most memorable example of contemporary urban landscape architecture.

When I ask my students how long they thought the project is, they can't say for sure, but some reckon up to fifteen kilometers.

In fact, the redesigned tranche of the High Line is just over two kilometers long. Considering the global prominence of these precious meters of urban infrastructure, we have to acknowledge that a well-designed, well-executed, comparatively compact project, can achieve colossal relevance. The story of the High Line is great because it demonstrates that the quality of urban public space can be successfully determined by its friends and users. Administrators and designers might have failed without the essential inspiration provided by those that live in the city. •

高线叙事

罗伯特·舍费尔

　　叙事，这一方法早已悄悄融入规划师的职业生涯中。它的意思就是，讲述一个故事。所谓叙事，应当塑造一个引人入胜的故事，精彩到足以流传或销售。故事的内容即是商品，必须交付到顾客手中。这一概念还有另一个英文中常用的市场营销术语——"讲故事"。在这种情况下，位于纽约肉类加工厂区的高线项目就如同装满故事的宝库，哪怕是马拉喀什城 Djemaa el-Fna① 集市上著名的说书人也会因为有它而兴奋许久。

　　西区牛仔是高线叙事中始终不变的部分。10 号大街有"死亡大道"之称，西区牛仔骑着马，走在货运列车前，以此警示行人注意安全。1941 年 3 月 29 日，21 岁的乔治·海德骑着他的"旋风"，护送了最后一班货运列车。这批货物居然不是肉制品，而是橙子。

　　2004 年，Field Operations 事务所与 Diller Scofidio + Renfro 事务所的建筑师一起，在高线公园的设计竞赛中获胜。詹

---

① Djemaa el-Fna，是摩洛哥的马拉喀什市最重要的集市广场，说书人汇集的地方。——译者注

姆斯·科纳展开了一个新的叙事："植-筑"①，就是把配置的植物单元与预制的混凝土铺装紧密地嵌套在一起，两者以多变的比例关系相互组合，塑造出新的高线公园特征。数百个不同的植物种类让这个线性公园独树一帜，成为北美最昂贵的城市花园；同时也让它成为纽约最大的旅游景点之一，切尔西地区收益最高的投资引擎。但是，只要在高线公园上欣赏著名的哈德逊河落日，就会不可避免地瞥见紧挨着的楼房的室内，比如Ten23 或 245Tenth 公寓。因此，已经有人指责高线公园像是一条窃听路线或者是偷窥者的公园。

植物，还为高线公园谱写了另外一个重要的叙事。2000年 3 月，乔尔·斯特菲尔德拍摄下盘踞在高线铁路半个世纪之久的野生植被，并出版了摄影集《漫步高线》②。那些总是在灰蒙蒙天空下拍摄的照片，浪漫得如此出人意料。它们对高线的保留、修复与设计重建起到了至关重要的作用。由皮特·奥多夫精心设计的植物，在帕特里克·库利纳的施工协助下，实现了詹姆斯·科纳的"植-筑"生态概念，但只是象征性地体现了一点点曾经昙花一现的荒野景象。

现在让我们回过头来，听一听原先那个让高线项目走向成功的故事。这个叙事的核心是意外、想象力、毅力和奉献。故事围绕着罗伯特·哈蒙德和约书亚·戴维展开，二人让雪球滚动了起来。他们先是阻止了本已决定的拆除方案，竭尽所

① "植-筑"概念，植被与建筑的结合。参见 Diller Scofidio + Renfro 官网介绍。——译者注
② 英文原著：Sternfeld, J. *Walking the High Line*（*First Edition*）. Steidl & Pace / MacGill Gallery, 2001. ——译者注

能地引起媒体的关注和参与；成立"高线之友"协会，筹集资金，并寻找到越来越多的支持者；举办设计竞赛，收到700多份参赛作品；在最终的实施方案竞赛中，詹姆斯·科纳和他的建筑师同事被选中为主创团队，打造了迄今为止最知名的风景园林项目，这是一个真正的成功故事！

Die High Line – ein Narrativ

Robert Schäfer

Das Narrativ hat sich eingeschlichen in unser Planerleben. Es bedeutet: eine Geschichte zu erzählen. Ein Narrativ soll eine spannende Geschichte formen, die erzählt oder vermarktet werden kann. Der Inhalt ist die Ware, die an die Konsumenten gebracht werden muss. Storytelling nennt man dies heute auf Englisch gerne im Marketingsprech. In diesem Kontext ist das Projekt High Line in New Yorks Meatpacking District eine Fundgrube für Geschichten, die die berühmten Märchenerzähler des Djemaa el-Fna in Marrakesch für lange Zeit glücklich machen würden.

Fester Bestandteil des High-Line-Narrativs sind die West Side Cowboys, die auf der 10., genannt Death Avenue, vor den Güterzügen her ritten, um andere Verkehrsteilnehmer zu warnen. Am 29. März 1941 begleitete der 21jährige George Hayde auf seinem Pferd *Cyclone* den letzten Güterzug auf der Straße. Die Fracht: nicht Fleisch, sondern Orangen.

Field Operations mit den Architekten von Diller Scofidio + Renfro gewannen den Wettbewerb für einen Park auf der stillgelegten Hochbahn 2004. James Corner führte ein neues Narrativ ein: agritecture. Natürliche und programmatische Gegebenheiten sollten in unterschiedlichem Verhältnis den Charakter der High Line prägen, also Betonelemente zum einen und Gräser, Stauden und Gehölze zum anderen. Hunderte von Pflanzenarten machen den linearen Park zu einem Unikum, zum teuersten Garten Nordamerikas – aber auch zu einer der größten Attraktionen der Stadt, zugleich zum effektivsten Investitionsmotor in Chelsea. Zu den gerühmten Aussichten auf den Hudson und die Sonnenuntergänge gesellen sich zwangsläufige Einblicke in die Wohnungen der dicht herangerückten Gebäude wie Ten23 oder 245 Tenth. Manche reden schon von der Pry Line, dem Voyeur Park quasi.

Die Pflanzen bilden eines der beiden wichtigsten Narrative der High Line. Joel Sternfeld fotografierte im März 2000 die Spontanvegetation, die sich auf der ein halbes Jahrhundert stillgelegenen Hochbahn angesiedelt hatte und brachte das Buch *Walking the High Line* heraus. Diese, stets bei grauem Himmel aufgenommenen, romantischen und völlig unverhofften Bilder trugen maßgeblich zum Erhalt der High Line und deren späteren Renovierung und Neugestaltung bei. Die meisterhaften Pflanzungen von Piet Oudolf, unterstützt von Patrick Cullina auf der Baustelle, bilden im Sinne von Corners *agri-texture* Ökosysteme nach, tragen jedoch das ephemere Ruderale nur noch im Geiste in sich.

Nun habe ich das Pferd von hinten aufgezäumt und komme

erst jetzt zum eigentlichen Narrativ des Erfolgsprojektes High Line. Hierbei geht es um Zufall, Phantasie, Beharrlichkeit, Hingabe. Es geht um Robert Hammond und Joshua David, die den Stein ins Rollen brachten, den schon beschlossenen Abriss der Hochbahn abwenden konnten, alle Register der Medienarbeit zogen, den Verein „Friends of the High Line" gründeten, Sponsorengelder eintrieben, immer mehr Unterstützer fanden, einen Ideenwettbewerb auslobten, der mehr als 700 Einsendungen brachte und schließlich den Realisierungswettbewerb, der James Corner und seinen Architektenkollegen die Urheberschaft an einem der bekanntesten Landschaftsarchitekturprojekte aller Zeiten brachte. Eine wahre Erfolgsgeschichte! •

城市自然主义者

弗雷德里克·施泰纳

对那些世界文化的探索者来说，高线公园使纽约更具吸引力。它在曼哈顿市中心引发了一场设计复兴，而这之前，詹姆斯·科纳就已经是一位颇具影响力的学者和具有煽动力的理论家。他的想法来源于他在宾夕法尼亚大学的导师、同事和竞争对手丰富的思想源泉——其中不乏倡导设计结合自然的伊恩·麦克哈格；推进城市生态学的安妮·惠斯顿·斯本；对建筑文化和实践保持深入参与的劳里·欧林。当你有这样的同伴时，每一个设计举措都会受到严格的审查，理性诚实也是必要的。

科纳的理论对设计中应用生态学的许多模糊之处提出了挑战，而他的实践则揭示了城市自然的巨大前景。这些前景在曼哈顿下城的高线——这个不太可能被废弃的铁轨上尤为明显。当然，高线公园并不是出自一位天才之手，但科纳所提出的卓越的愿景，提升了这个项目的方方面面。项目的确亦有先例：特别是巴黎林荫道和西雅图煤气厂公园。但是，就像曼哈顿的其他标志性景观，弗雷德·劳·奥姆斯特德和卡尔弗特·沃克斯的中央公园一样，高线公园在野心和影响力方面都使其前辈黯然失色。像中央公园一样，各个城市的领导者都想拥有自己的高线公园。突然之间，被遗失的城市空间得以重现。

与中央公园一样，一种新的景观美学应运而生，这与科纳对雷姆·库哈斯和斯坦·艾伦等设计理论家的深入解读密切相关。科纳的作品"从关于场所、人和自然的生态学中获取信息和灵感"。伴随着高线公园的建成，人、环境与其他生物之间产生了新的互动。

在任何季节，沿着高线公园散步都是一种乐趣，在不断变化的城市节奏中，唯有这条路线始终如一。作为一个从高处看到的景象和一个庇护地，这条步道所提供的城市景观是其他任何地方都无法找到的。由大师皮特·奥多夫挑选的植物在科纳及其团队策划的广告牌和街景中竞相吸引着游客的注意力。在高空漫步时，我们四处环顾，内心也随之打开了。

高线公园将我们与全人类联系起来，并将我们推向人类历史的舞台。我们与环绕在周围的其他物质相连：海风、晨露和集会的鸟儿。景观设计的挑战包括如何设想远处的林中空地，以及如何勾勒出无法感知的变化。自然依规律运行（例如重力），但却在非凡的复杂性中演变。

詹姆斯·科纳、Field Operations 事务所的高线公园及其后续项目都重新定义了风景园林的发展轨迹和潜力。在 21 世纪城市未来的讨论和计划中，科纳所展现的领导力将风景园林置于中心位置。高线公园说明了安全又美丽的城市景观是如何促进人类和经济的健康发展。同时，一条废弃廊道的恢复和改造也提供了希望与灵感。

The Urbane Naturalist

Frederick Steiner

Before the High Line made New York City even more irre-
sistible to the world's culture seekers and sparked a design renais-
sance in downtown Manhattan, James Corner was an influential
academic and provocative theorist. His ideas emerged from the
rich seedbank promulgated by his mentors, colleagues, and rivals
at the University of Pennsylvania—none other than Ian McHarg,
who advocated designing with nature; Anne Whiston Spirn, who
advanced urban ecology; and Laurie Olin, who maintained a
deep involvement in architectural culture and practice. When you
keep this kind of company, every design move is subject to intense
scrutiny, and intellectual honesty isn't optional.

Whereas Corner's theories challenged many of the ambigu-
ities of applying ecology within design, his practice reveals the
great prospects for urban nature. These prospects are especially
apparent in the unlikely abandoned rail site of the High Line in
lower Manhattan. Of course, the High Line is not the result of a
single genius, but the brilliance of Corner's vision elevates every

aspect of the project. Of course, there were precedents: notably *Promenade Plantée* in Paris and *Gas Works Park* in Seattle. However, like Manhattan's other iconic landscape, Frederick Law Olmsted and Calvert Vaux's *Central Park*, the High Line eclipsed its predecessors in both ambition and influence. Like Central Park, city leaders everywhere wanted their own High Line. Suddenly lost urban space was found.

As with Central Park, a new landscape aesthetic emerged, one closely linked to Corner's deep readings of design theorists such as Rem Koolhaas and Stan Allen. Corner's work is "informed and inspired by the ecologies of place, people, and nature." With the High Line, new interactions between people and their environments and other living organisms have been created.

A walk along the High Line is a joy in any season, a fixed course though an ever-changing sea of urbanity. A prospect and refuge, the walkway offers views of the city not to be found anywhere else. The plants, selected by the master Piet Oudolf, compete for the flâneur's attention with the billboards and street scenes in found landscapes curated by Corner and his team. Our minds act like street photographers as we stroll aloft.

The High Line connects us to humankind and puts us on the stage of human history. We are connected to the other physical stuff that surrounds us outdoors: an ocean breeze, the wetness of the morning dew, a convention of birds. The challenges of landscape design include how to envision that clearing in the distance and how to frame the imperceptibility of changes. Nature plays by laws—gravity, for instance—but plays out in marvelous complexity.

James Corner, the High Line, and the subsequent work of Field Operations have redefined the trajectory and the potential of landscape architecture. Corner's leadership has placed landscape architecture center stage in discussions and plans for the future of the 21st-century city. The High Line illustrates how safe, beautiful, urban landscapes can promote human and economic health. The recovery and transformation of an abandoned corridor provides hope and inspiration. •

转变

安杰·斯托克曼

　　我认为，高线公园是一个具有突破性意义的项目，因为它的出现标志着我们正处于一场历史变革之中。我们已经越来越清晰地意识到，这是一场风景园林师对自我认知的转变。

　　从工业化开始，人与自然之间的关系就发生了翻天覆地的变化。高线铁路这样的大型工程设施就是工业社会的遗迹。工业革命之后，人类利用大量的能源和资源对自然进行了一场史无前例的改造。城市的急剧扩张被看作是破坏自然景观的元凶，而人类自己也被认为是在人类世时代改变地球的绝对力量。此后，社会上的主流观点是：建筑物与自然之间是一种非黑即白的对立关系——"建造是邪恶的""自然是美好的"。一边是由建筑师和工程师建造的，用技术堆砌的城市；另一边则是尚未开发的开放空间，确切地说，是风景园林师与规划师保护的自然景观遗迹。

　　高线公园的案例表明：转化并再利用旧工业时代的建筑和场地，可以为景观基础设施设计以及城市发展提供意料之外的可能性。荒废的铁轨、采石场、机场用地和工业废弃地，还有闲置的屋顶与复垦的垃圾山都是城市景观重构的重要空间资源，是一种令人振奋的挑战。对此，高线公园的设计体现了一种看待事物的新角度——不再划分自然、基础设施和建筑之间的界限。自然与建筑不再是对立的关系，景观也不再是尚未开发的开放空间。高架铁路转变为公园，促使了动态景观元素、技术系统与人类生活之间紧密联系的建立。

风景园林学科将打破几个世纪以来的固化关系，推动建造原则的转变，从曾经的忽视自然，转变为建筑与自然以及景观资源相结合。随着世界各地越来越多的人生活在城市聚集区，高线公园这样的项目具有特别的示范意义，它实验性地展示了人与自然之间的积极关系。每个人都像"建筑师"一样，作为改变地球的力量，设计并塑造了自然。风景园林师将建筑与景观看作是"人造"与"自然"的混合体，并在此基础上进行设计和建造——这是风景园林学科在人类世时代的前瞻性表现。

Kurswechsel

Antje Stokman

Die High Line ist für mich ein wegweisendes Projekt, da sie in ikonischer Weise dafür steht, dass wir inmitten eines historischen Umbruchs stecken – dessen Konturen sich immer deutlicher abzeichnen und der einen Kurswechsel für unser Selbstverständnis als Landschaftsarchitekten bedeutet.

Das menschliche Verhältnis zur Natur hat sich seit Beginn der Industrialisierung in dramatischer Weise verändert: Ingenieurbauwerke wie die High Line sind Relikte unserer Industriegesellschaft, die seit Beginn der industriellen Revolution unter Einsatz enormer Materialund Energieressourcen eine beispiellose Umformung von Natur mit sich gebracht haben. Das extreme Wachstum der Städte wirkte als Motor der Landschaftszerstörung und führte zu einer Positionierung des Menschen im Anthropozän als maßgebliche erdverändernde Kraft. Seitdem werden die gesellschaftlichen Vorstellungen zum Verhältnis zwischen Gebautem und Natur beherrscht vom kategorialen Gegensatz zwischen dem „bösen Bauen" und der „guten Natur": Auf der einen Seite die durch Architekten und Ingenieure gebaute, technisch überformte Stadt – auf der anderen Seite die Gestaltung der nicht bebauten Freiräume bzw. Unterschutzstellung naturlandschaftlicher Relikte durch Landschaftsarchitekten und -planer.

Wie die High Line zeigt, bietet die Transformation der gebauten Strukturen des Industriezeitalters ungeahnte Möglichkeit für die Entwicklung und Gestaltung einer landschaftlichen Infrastruktur für unsere Städte. Aufgegebene Gleistrassen, Kiesgruben, Flughafenareale und Industriebrachen, ungenutzte Dächer und rekultivierte Müllberge stellen eine spannende Herausforderung und wichtige Raumressource für die Neustrukturierung städtischer Landschaften dar. Dabei steht die Gestaltung der High Line sinnbildlich für eine neue Sichtweise, die die Unterscheidung zwischen Natur, Infrastruktur und Architektur aufhebt. Natur ist nicht mehr das Gegenüber von Architektur, Landschaft ist nicht mehr der unbebaute Freiraum. Die Transformation der High Line von einer Gleistrasse zum Park ermöglicht eine spannungsvolle Beziehung zwischen dynamischen Landschaftselementen, technischen Systemen und menschlichen Lebenswelten.

Damit wird die Landschaftsarchitektur zum führenden Impulsgeber für die Abkehr vom naturvergessenen Prinzip des Bauens hin zur Einbettung von Bauten in natürliche, landschaftliche und stoffliche Ressourcenzusammenhänge, wo einige Jahrhunderte lang die Abkopplung von diesen im Vordergrund stand. Da Menschen weltweit zunehmend in urbanen Agglomerationen leben, haben Projekte wie die High Line als Experimentierfelder und Schaufenster für die exemplarische Demonstration von positiven Verbindungen zwischen Mensch und Natur eine herausragende Stellung. Der Mensch als erdverändernde Kraft formt und gestaltet die gesamte Natur, genau wie er Architekturen formt und gestaltet. Auf dieser Basis können bauliche Eingriffe im Sinne einer produktiven Durchdringung von „menschgemachter" und „natürlicher" Architektur und Landschaft gedacht, entworfen und realisiert werden – als zukunftsweisender Ausdruck einer Landschaftsarchitektur im Anthropozän. •

成功是迷人的

迪特玛·斯特劳布

纽约的建筑似乎直冲云霄，整个世界都想变得和纽约一样酷。其他城市也试图复制"曼哈顿模式"，但纽约却总是设法领先一步。随着高线公园的发展，城市不再照搬其垂直发展模式，并创造了一种彻底的水平移动方式。

作为一名设计师，你总是抱着这样的希望：有朝一日塑造一个极具风格且有着强大光环的空间，一个某种程度上散发着勇气和胆量的地方，你希望人们能够体验到这种空间感受。但你必须准备好抓住这些稍纵即逝的机会，而詹姆斯·科纳就抓住了摆在他面前的机会。

高线公园不是引领潮流的，它是前卫的。潮流只是引发了由个人或市场产生的短暂现象。源自前卫思想家的推动力和转变更具根本性，它们对设计、美学和环境伦理产生更长久的影响。前卫派的高线公园设计鼓励一种先锋性作用，激发了具有开创性的视角与想法。

十多年来，我一直作为一名风景园林师在加拿大温尼伯工作。我的建筑师同事在参考纽约高线公园时表达了他们对"景观"的尊重。这种时髦的引用是为了创造相互理解，但它也凸显了该行业华丽的词藻与日常实践之间一直存在的差距。然而，高线公园的最大作用是它将空间设计师这一群体联系起来。毫无疑问，高线公园强化了风景园林学科的声誉和身份，在我看来，它是一次无可争议的成功。

这一伟大成就也产生了巨大的风险，可以说成功是迷人的。高线公园项目之后，出现了由明星建筑师设计的诸多豪华塔楼和气派的建筑。在重新估值的城市地区，绅士化进程达到巅峰。只有少数原来的屠夫和肉商仍在这里为该地区供应肉制品和保存这一地区的烟火气。"屠宰"将继续下去，而房地产市场则沉迷于这块上等的"肥肉"。

我很想参观纽约的建筑、公园和广场。我想漫步于三层楼高的高线公园之中。我期待着与青草、鲜花相遇，欣赏美景。我希望高线公园一直是一个栖息地，能让漫步者和环球旅行者，热情购物的人和筋疲力尽的商人，银行家和妓女，当地人和好奇的游客，乞丐和小偷……都可以在如此蓬勃多样的城市氛围中徘徊。

Success is Sexy

Dietmar Straub

The whole world wants to be as cool as New York with its buildings that seem to grow straight into the sky. Other cities try to copy the "Manhattan Model", but New York manages to always stay one step ahead. With the development of the High Line, the city refused to copy its own model of vertical mobility and created a radical horizontal movement.

As a designer, you always hold on to the hope that you will one day shape a space with a strong essence and powerful aura, a place that exudes some degree of courage and audaciousness which you hope people will experience as a physical sensation. But you have to be prepared to jump at these fleeting chances, and James Corner snatched his opportunity.

The High Line is not trendsetting, the High Line is avant-garde. Trendsetting merely initiates short-term phenomena generated by individuals or the market. The impulses and transformations resulting from avant-garde thinkers are more fundamental and achieve a greater long-term impact on design, aesthetics and environmental ethics. The avantgardist's High Line design has instigated a pioneering role and stimulated groundbreaking perspectives and ideas.

For more than a decade now I have worked in Winnipeg, Canada as a landscape architect. My architect colleagues show their respect for "landscape" by referencing the High Line in New York. This trendy citation is intended to create mutual understanding. However, it highlights the persistent gap between the profession's rhetoric and the reality of everyday practice. Nevertheless, the great merit of the High Line is that it works, as the paving sand to tie the community of spatial designers together. Without a doubt, the High Line strengthens the reputation and identity of landscape architecture and is, in my opinion, an undisputed success.

This great achievement also generates a significant risk - success is sexy. The High Line project has been followed by many luxury towers and prestige buildings designed by star architects. Processes of gentrification have reached their peak in the revalued urban area. Only a few of the original butchers and meat traders are still there supplying meat products and authenticity for the district. The "slaughtering" is going to continue. The real estate market is addicted to prime ribs.

I would love to visit New York's buildings, parks and squares. I want to walk the High Line hovering three stories above the city's ground. I look forward to encountering the grasses, the flowers, and enjoying the bella vistas. I hope the High Line remains a habitat for the flâneurs and the globetrotters, the passionate shoppers and stressed-out business people, the bankers and the prostitutes, the locals and the inquisitive tourists, the beggars and the pickpockets... to linger in this state of thriving urban diversity. •

好奇的园丁

安娜·瑟梅尔

在漫长的 6 个月里，大概是加拿大的冬季，一个好奇的小男孩莱姆勇敢地寻求着将一条乡村铁路改造成一个欣欣向荣的花园，这个故事吸引了我的孩子们的注意力和想象力。莱姆把一座灰暗的大城市改造成了一个绿意盎然的地方，男孩付出的艰苦卓绝的努力令他们着迷，他谦逊和好奇的天性让孩子们欣然入眠。

彼得·布朗的《神奇花园》① 于 2009 年出版。莱姆发现了一个黑暗的楼梯，它通向一条旧铁路线，一个被遗忘的世界。他首先注意到的是一片色彩斑斓的、却即将枯死的野花。莱姆毫不犹豫地成为一名园丁。然而，他对如何照料植物和花园并没有经验，但他凭借直觉知道这里需要他的帮助。他浇水、修剪并倾听花草的声音，花园不断扩大，将这座沉闷的城市重新装点成美丽的家园。

曼哈顿高线公园项目的故事也源于一个小范围的地方倡议，随着时间的推移它逐渐发展起来。两位当地的倡导者成立

---

① 英文原著：Brown, P. *The Curious Garden*. Little Brown Books for Young Readers, 2009.——译者注

了"高线之友"协会，他们支持对铁路的再利用和保护。一座生锈的高架铁路被改造为一个充满活力的公共空间，这座茂盛的花园蜿蜒盘旋在纽约繁忙街道的上空。许多满怀热忱的新晋园丁出人意料地伸出援手，进一步强化了这一景观，证明了该项目从一开始就扎根基层。

我们曾问过，如果没有中央公园，纽约会变成什么样子？现在我们可以问，如果没有高线公园，纽约又会变成什么样子？而且，若没有那些好奇的园丁，花园又会是什么样子？"高线之友"协会成功地筹集了数百万美元，并持续募集资金，招募当地志愿者，使这一富有感染力的公园理想保持生机。

过去，如果没有国王的命令，勒诺特尔也无法建造出世界上最著名、最有影响力的花园。如今，社区参与景观项目的规划过程已几乎成为标准。在需求方面，满怀热情的个体与公共利益相关者一起取代了传统客户。此外，曼哈顿高线公园的改建唤起了更高的社区参与度，那些热心市民自发投入时间和金钱，因为绝大部分资金无法由政府或市政当局提供。这个著名的项目标志着参与式设计过程的一个转折点，让它超越了普通的公共研讨会和听证会。

我们可以从这一杰作中学到的是，风景园林师的角色正面临着转型所带来的挑战。社区的支持是必要的，也是必须的，这样才能将设计师富有想象力的图纸变成现实的设计作品。它要求精于此道的专业人士利用他们的智谋和批判性思维，与公众一起为人们创造切实可行的想法。

　　有一点是始终正确的，园艺仍在唤起人们对新发现的喜悦。自然仍然可以在最不可能的地方茁壮成长。事实上，莱姆和高线公园所传递出的希望也许能克服人造世界的许多问题。

The Curious Gardeners

Anna Thurmayr

For six long months, about the time of a Canadian Winter, the story of an inquisitive little boy named Liam and his courageous quest to transform a rustic railway into a flourishing garden captured my children's attention and imagination. They were fascinated by Liam's heroic efforts to transform a big grey city into a lush green place and were lulled to sleep by the boy's humble and curious nature.

The Curious Garden by Peter Brown was published in 2009. Liam discovers a dark stairway leading up to a forgotten world of an old rail line. The first thing he notices is a coloured patch of dying wildflowers. Without hesitation Liam becomes a gardener. He has no experience in how to tend to plants and gardens, yet he intuitively knows his help is needed. As he waters and prunes, listening to the grasses and flowers, the garden expands to redecorate the dull city into a beautiful habitat.

The story of Manhattan's High Line project also began with a small local initiative that grew over time. Two local advocates founded the "Friends of the High Line" group and championed the railway's reuse and preservation. The transformation of the rusty railway viaduct resulted in a vibrant public space with a thriving garden flowing and meandering above New York's busy streets. A testament to its grassroots beginnings, the landscape was further enhanced by the many new and eager gardeners who unexpectedly pitched in to help.

We used to ask what would New York be without its *Central Park*? Now we can ask what would New York be without its High Line? And further, what is a garden without its curious gardeners? "Friends of the High Line" successfully raised several million dollars and continues to solicit capital and local volunteer efforts to keep this infectious ideal alive.

In the past, Le Nôtre would have been unable to create the world's most celebrated and influential gardens without the king's orders. Today, however, community involvement in the planning process of a landscape project has become almost standard. Passionate private individuals together with public stakeholders have replaced the traditional client's orders. What's more, the conversion of Manhattan's High Line invoked an even higher communal engagement, and these caring citizens still invest their own time and money, as the vast majority of funding does not come from government or municipalities. This prestigious project marks a turning point in participatory design process and goes far beyond the standard public workshops and hearings.

What we can learn from this masterpiece is that the role of the landscape architect is also being challenged to transform. Community support is necessary and imperative to turn a designer's imaginative drawings into realized designs. It calls for well-versed professionals to use their resourcefulness and critical thinking skills in order to create tangible ideas not just *for* people, but *with* them as well.

One thing remains true. Gardening still evokes the delight of new discoveries. Nature can still thrive in the most unlikely of places. And the hopeful message of Liam and the High Line may in fact triumph over many problems of the man-made world. •

走向第三条路

克里斯蒂安·沃特曼

　　毫无疑问，詹姆斯·科纳是我们这个时代最具影响力的风景园林理论家和杰出的实践者之一。1996 年，当他的开创性著作《测量美国景观》问世时，他极佳的拼贴作品与亚历克斯·麦克林的航拍相结合，揭示了景观是一个在不同时间维度上蕴含着复杂动态变化的层次丰富的实体。曾有这样一份明确的声明（即使不是一个宣言）：风景园林师的研究对象是如此的复杂，以至于传统的景观描述手段是远远不够的，甚至削弱了我们的转化潜力。

　　在 20 世纪 90 年代关于景观的讨论中，景观规划和风景园林之间的关系逐渐疏远，科纳却在其中开辟了第三条极其重要的景观探索之路。当新兴的景观规划领域基于自然科学的严格方法找到自己的立足点时，其"兄长"风景园林正处于围绕艺术进行探索的尾声。在随后的诸多探索中，科纳和越来越多的专业人士描绘出另一条路，解决了大中型城市景观塑造过程混乱的问题，而这些问题在当时无论是风景园林还是景观规划都没能很好地给予回答。在 21 世纪的第一个十年中，当"景观都市主义"的理念发展到一个可建造的尺度时，这些想法便开始在实践中得到检验，其中最重要的可能就是科纳自身实践中的大量委托项目。随着第一批项目开始产生影响，特别是在当斯维尔公园和清泉垃圾填埋场公园竞赛之后，许多人开始更好地理解"景观都市主义"的叙事价值，而高线公园项目则带来了另外一系列问题。

在科纳的早期著作中，他的清泉垃圾填埋场竞赛作品通常被视为一次振奋人心的尝试，而高线公园往往被认为是一个更符合传统风景园林范围的设计项目。从表面上看，人们可以推测高线项目受限于预算、完成期限和公众对于"成品"的期望。但仔细观察，人们发现科纳对这两个截然不同的项目的构想并不矛盾。

在他最近编著的《景观之想象》①一书中，科纳承认他的文章是"一名风景园林师的临时猜想，他主要对设计和从事实际项目感兴趣，但同时，他也在为这项工作寻求真谛以及它所具备的更广泛的文化意义"②。在一生的追求中，科纳成为一名探究国土空间的知名专家。在学术界之外，科纳不得不在一个由市场力量、公众期望和政治意愿组成的延展的生态系统中寻找方向，并一再拓展他的整套设计策略。他对"渴望的想象，场所营造的诗意，以及设计的实体（塑造着一个新的充满活力

---

① 英文原著：Corner, J., Hirsch, B. A. *The Landscape Imagination*：*The Collected Essays of James Corner 1990-2010*. Princeton Architectural Press, 2010. ——译者注
② 詹姆斯·科纳，《景观之想象：詹姆斯·科纳思想文集》。詹姆斯·科纳，艾利森·赫希（编）普林斯顿建筑出版社，2010. p. 7.

的公共领域)"①，这些兴趣都在高线公园中得到了最突出的体现，这一项目也是在文化层面对风景园林的又一次出色的调查和研究。

---

① 详见 2014 年珍妮特·索迪对詹姆斯·科纳的采访。收录于珍妮特·索迪，《超越都市主义》。LISt 实验室，2014. p. 127.

Walking the Third Path

Christian Werthmann

James Corner is doubtlessly one of the most influential theo-rists and brilliant practitioners of landscape architecture in our time. When his seminal book *Taking Measures Across the American Landscape* came out in 1996, his splendid collages in combination with the aerial photography of Alex MacLean revealed landscapes as richly layered entities engulfed in complex dynamics on dif-ferent time scales. It was a clear statement, if not manifesto, that landscape architects are engaged in a subject that is so complex that its traditional means of description are not only utterly inadequate, but cripple our transformative potential.

In the discussion about landscape in the 90s, Corner opened up a desperately needed third path of investigation next to the es-tranged siblings of landscape planning and landscape architecture. While the still emerging field of landscape planning was finding its footing through a rigid approach based in the natural sciences, its "older" brother, landscape architecture was at the tail-end of an exploration revolving around art. The ensuing explorations by

Corner and a growing group of professionals charted an alternate path, addressing questions surrounding the messy processes of medium- to large-scale urban landscapes, which at the time neither landscape architecture nor landscape planning were well equipped to answer. In the first decade of the new millennium the ideas of *Landscape Urbanism* began to be tested in practice when they moved to a buildable scale, perhaps most significantly with the large commissions of Corner's own practice. As the first projects began making an impact, particularly after the *Downsview Park* and *Fresh Kills* competitions, many began to better understand the value of the *Landscape Urbanism* narrative, while the High Line project introduced another set of questions.

As Corner's entry to the *Fresh Kills* competition has generally been seen as an exciting testing ground for his earlier writing, the High Line is often perceived as a project more in line with the traditional scope of landscape architecture. On the surface one could speculate that Corner in his High Line project had succumbed to the constraints of budgets, deadlines, and the public desire for a "finished" product. But on a closer look one finds no contradiction in Corner's conception of these two very different projects.

In his recently co-edited book *The Landscape Imagination*, Corner admits that his essays are "the provisional conjectures of a landscape architect, who is primarily interested in designing and making actual projects, but who is, at the same time, searching for a deeper *raison d'être* and broader cultural relevance for that work." [1] In this lifelong pursuit, Corner became an expert master in understanding territory. Outside of academia, being forced to navigate an expanded ecosystem of market forces, public desires, and political will, Corner has been expanding his set of strategies. His interest in "the imagination of desire, the poetics of place-making, and the physicality of design in forging a freshly vibrant public realm" [2] was most prominently revealed in the High Line project. It stands as a remarkable re-investigation and re-investment into the cultural side of landscape architecture. •

---

[1] James Corner, *The Landscape Imagination: The Collected Essays of James Corner.* James Corner and Alison Hirsch, eds. Princeton Architectural Press, 2010. p. 7.

[2] see Jeannette Sordi's 2014 interview with James Corner. In: Jeannette Sordi, *Beyond Urbanism.* LISt Lab Laboratorio. 2014. p. 127.

科纳的新景观宣言

俞孔坚

　　三次参观该项目后，我将高线公园解读为对詹姆斯·科纳《新景观宣言》[1]的一种体现。2016 年 6 月，他在宾夕法尼亚大学明确提出这一宣言，当时正值伊恩·麦克哈格等人签署《关注宣言》[2]（1966 年）50 周年庆典之际。科纳声称："我们承担着自己的使命——一个新的宣言，即从定量与定性、生态与社会、实用与诗意的角度，景观设计师必须接受塑造与构建未来城市的挑战"。高线公园象征着一种回归或是一次收复，使城市重新成为景观设计专业的主战场。正如《关注宣言》所宣称的那样，自 20 世纪 60 年代以来，由于发达国家对环境问题的关注，景观设计专业已从城市和私人住宅中的观赏园艺以及城市重建与发展，转向生态规划和土地管理这一新领域。这种转变使该行业在治愈被工业化和城市化摧毁的土地方面发挥了前所未有的新领袖作用。然而，它也冒着使这些景观项目趋于平庸而不为人知的风险，并否定了景观通过我们的艺

---

① 英文原著：*New Landscape Declaration*. ——译者注
② 英文原著：*A Declaration of Concern*, June 1 and 2, 1996. https://www.lafoundation.org/who-we-are/values/declaration-of-concern. ——译者注

术来定义和加强人类身份。目前，在北美洲和欧洲，如果无法从根本上解决全球气候变化和其他方面的问题，那么环境危机至少已在当地的具体领域得以克服。事实上，这场危机已经转移到发展中国家，并且日益严重。至少在詹姆斯·科纳和Field Operations 事务所的例子中，景观设计专业将后工业城市作为战斗的主要前沿阵地；景观不仅被视作一种美化和点缀城市的艺术品，而且作为一种都市主义的方式。这种态度意味着"把城市看作花园"，旨在将"经验性和实用性提升到诗歌和艺术的高度"①。同时，以高线公园为例，科纳的作品有着深厚的社会、经济与文化基础，在其他项目中也有着扎实的生态学基础。

　　高线公园的成功不仅激励了发达国家后工业城市的发展，而且对那些正忙于新城建设并遭受饮用水短缺、空气与水严重污染和种族隔离的国家来说，高线公园也是一盏指路明灯。它的成功表明，如何从一开始就将景观作为基础设施进行规划和设计，从而形成一个更美好的未来城市——基础设施可以为城市整合各项服务，包括行人的安全通行、为生物多样性提供生命支持、促进多元化社会的文化和社会融合，以及娱乐和审美体验。该项目获誉无数，它不仅恢复了城市中那些曾经几乎失去的部分，而且使景观设计专业重拾荣光，正如当年奥姆斯特德的中央公园展示的这一行业对城市尺度运作的潜在影响。它还向世界证明，景观也许是将社会、文

---

① 詹姆斯·科纳，《新景观宣言》，风景园林基金会（LAF），2017. p. 67.

化、生态或经济等多种过程结合在一起的唯一媒介，这使得
景观成为一种真正的都市主义艺术，能够创造出深层的城市
形态。

A Manifesto for Corner's New Landscape Declaration

Kongjian Yu

Having visited the project three times, I read the High Line as a manifestation of James Corner's *New Landscape Declaration* which he has made so clearly in June 2016 at the University of Pennsylvania, on the occasion of the 50th anniversary celebration of the *Declaration of Concern* signed by Ian McHarg and others in 1966. Corner declared: "Here we have our mandate — a new declaration that landscape architects must take on the challenges of shaping and forming the future city, quantitatively and qualitatively, ecologically and socially, pragmatically and poetically". The High Line symbolizes a return to, or rather a reclaim of, the city as the main battlefield for the profession of landscape architecture. The environmental concern that has grown in the developed world since the 1960s has drawn the profession of landscape architecture away from ornamental gardening in the cities and private estates, city rebuilding and urban development towards the frontier of ecological planning and land stewardship, as declared by the *Declaration of Concern*. This shift has given the profession an unprecedented new leadership role in healing land devastated by industrialization and urbanization. Yet, it has also risked allowing such landscape architecture projects to become invisible, negating the role such interventions might have in defining and reinforcing human identity through our art. Now, in North America and Europe the environmental crisis has been overcome at least locally in the tangible sphere, if not fundamentally with regard to global climate change and beyond. As a matter of fact, though, that crisis has shifted towards developing countries and it has become even worse. The profession of landscape architecture — at least in the case of James Corner and Field Operations — takes the post-industrial city as the main frontier of the battle; this time not only as an art of beautifying and making embellishing artifacts in the city, but as a way of urbanism. This attitude means taking the "city as a garden" in order to "[elevate] experience and pragmatics to poetry and art"[①]. At the same time, Corner's work is deeply grounded socially, economically, and culturally in the case of High Line, and also ecologically in some of his other projects.

The success of the High Line not only inspired the post-industrial cities in the developed world, it is also a glowing beacon to countries now busy with building new cities and suffering with the shortage of potable water; with heavy air and water pollution

---

① James Corner, *The New Landscape Declaration*. Landscape Architecture Foundation (LAF), 2017. p. 67.

and with a segregated population. The High Line's success signified how a better future city can be formed through planning and designing landscape as infrastructure from the outset — and that infrastructure can integrate various services for the city, including safe movement for pedestrians, life support for biodiversity, social and cultural integration for a diverse society, alongside recreation and aesthetic experience. The acclaim the project has received has not only reclaimed part of the city that was almost lost, but it has also made the profession of landscape architecture itself at least as visible as when Olmsted's *Central Park* demonstrated the potentially city-scale operations of the profession. It also demonstrated to the world that landscape is perhaps the only medium that unites so many kinds of process, be it social, cultural, ecological or economic, which makes landscape architecture a true art of urbanism that capable of creating deep urban forms. •

作者简介

托比约恩 · 安德森

风景园林学 教授

瑞典农业科学大学，瑞典乌普萨拉

事务所：Sweco 建筑事务所，瑞典斯德哥尔摩

阿德里 · 范 · 登 · 布林克 博士

风景园林与土地利用规划 教授

瓦赫宁根大学，荷兰

保罗 · 比尔吉

风景园林学 教授

宾夕法尼亚大学，美国费城

威尼斯建筑大学，意大利

米兰理工大学，意大利

事务所：Bürgi 工作室，瑞士卡莫里诺

米歇尔·德斯维涅

凡尔赛国立高等园艺学院 院长，法国

事务所：MDP 事务所，法国巴黎

约格·德特马 博士

开发空间规划与设计教席 教授

达姆施塔特技术大学，德国

索尼娅·邓佩尔曼 博士

风景园林学 教授

哈佛大学，美国

马塞拉·伊顿 博士

风景园林学 教授

马尼托巴大学，加拿大温尼伯

阿德里安·高伊策

风景园林学 教授

瓦赫宁根大学，荷兰

事务所：West 8，荷兰鹿特丹

克里斯托夫·吉罗特

风景园林学 教授

苏黎世联邦理工学院，瑞士

事务所：Girot 工作室，瑞士戈克豪森

沃尔夫拉姆·霍夫 博士

风景园林学 教授

罗格斯 – 新泽西州立大学，新不伦瑞克

约翰·迪克森·亨特 博士

景观历史和理论 名誉教授

宾夕法尼亚大学，美国费城

安德烈亚斯·基帕尔 博士

风景园林学 教授

热那亚大学，意大利

米兰理工大学，意大利

事务所：LAND Srl，意大利米兰；KLA 景观事务所，德国杜塞尔多夫

彼得·拉茨

风景园林设计与规划系 名誉教授

慕尼黑工业大学，德国

事务所：Latz + Partner，德国克兰茨贝格

戴维·莱瑟巴罗 博士

建筑学 教授

宾夕法尼亚大学，美国费城

莉莉·利卡

风景园林学 教授

维也纳自然资源与生命科学大学，奥地利

事务所：koselička 工作室，奥地利维也纳

弗兰克·罗尔伯格 博士

风景园林学 教授

亚琛工业大学，德国

事务所：Lohrberg 城市景观设计事务所，德国斯图加特

瓦莱里奥·莫拉比托 博士

风景园林学 教授

雷焦卡拉布里亚地中海大学，意大利

宾夕法尼亚大学，美国费城

事务所：APscape 工作室，意大利雷焦卡拉布里亚

乔·努内斯

风景园林学 教授

里斯本大学，葡萄牙

卢加诺大学，瑞士门德里西奥

事务所：PROAP 风景园林设计研究室，葡萄牙里斯本

约格·雷基特克 博士

风景园林学 教授

墨尔本皇家理工学院（RMIT）/ 澳大利亚

罗伯特·舍费尔

Landezine 网站风景园林专业编辑

Garten + Landschaft 期刊前主编

TOPOS 风景园林与城市设计期刊的国际专栏作者

弗雷德里克·施泰纳 博士

佩利教授和设计学院 院长

宾夕法尼亚大学，美国费城

安杰·斯托克曼

风景园林学 教授

汉堡港口城市大学（HCU），德国汉堡

迪特玛·斯特劳布

风景园林学 教授

马尼托巴大学，加拿大温尼伯

阿兰·塔特 博士

风景园林学 教授

马尼托巴大学，加拿大温尼伯

安娜·瑟梅尔

风景园林学 教授

马尼托巴大学，加拿大温尼伯

乌多·维拉赫 博士

风景园林与工业景观教席 教授

慕尼黑工业大学，德国

克里斯蒂安·沃特曼

风景园林设计与设计教席 教授

汉诺威大学，德国

俞孔坚 博士

景观设计学 教授

北京大学建筑系，中国

事务所：土人景观，中国

Autorinnen und Autoren

Thorbjörn Andersson

Professor für Landschaftsarchitektur

Sveriges lantbruksuniversitet (SLU), Uppsala/ Schweden

Büro: Sweco Architects, Stockholm/ Schweden

Adri van den Brink, Dr.

Professor für Landschaftsarchitektur und Landnutzungsplanung

Wageningen University & Research (WUR) / Niederlande

Paolo Bürgi

Professor für Landschaftsarchitektur

University of Pennsylvania, Philadelphia/ USA,

Università Iuav di Venezia/ Italien,

Politechnico di Milano (POLIMI) / Italien

Büro: Studio Bürgi, Camorino/ Schweiz

Michel Desvigne

Präsident der École nationale supérieure de paysage

de Versailles (ENSP) / Frankreich

Büro: MDP Michel Desvigne Paysagiste, Paris/ Frankreich

Jörg Dettmar, Dr.

Professor für Entwerfen und Freiraumplanung

Technische Universität Darmstadt/ Deutschland

Sonja Dümpelmann, Dr.

Professorin für Landschaftsarchitektur

Harvard University/ USA

Marcella Eaton, Dr.

Professorin für Landschaftsarchitektur

University of Manitoba, Winnipeg/ Kanada

Adriaan Geuze

Professor für Landschaftsarchitektur

Wageningen University & Research (WUR) / Niederlande

Büro: West 8, Rotterdam/ Niederlande

Christophe Girot

Professor für Landschaftsarchitektur

ETH Zürich/ Schweiz

Büro: Atelier Girot, Gockhausen/ Schweiz

Wolfram Höfer, Dr.

Professor für Landschaftsarchitektur

Rutgers – The State University of New Jersey, New Brunswick/ USA

John Dixon Hunt, Dr.

Professor Emeritus für Geschichte und Theorie der Landschaft

University of Pennsylvania, Philadelphia/ USA

Andreas Kipar, Dr.

Professor für Landschaftsarchitektur

Università degli studi di Genova/ Italien

Politechnico di Milano (POLIMI) / Italien

Büro: LAND Srl Mailand und KLA kiparlandschaftsarchitekten

Milano, Düsseldorf/ Italien, Deutschland

Peter Latz

Professor Emeritus für Landschaftsarchitektur und Planung

Technische Universität München/ Deutschland

Büro: Latz + Partner, Kranzberg/ Deutschland

David Leatherbarrow, Dr.

Professor für Architektur

University of Pennsylvania, Philadelphia/ USA

Lilli Lička

Professorin für Landschaftsarchitektur

Universität für Bodenkultur (BOKU) Wien/ Österreich

Büro: koselička, Wien/ Österreich

Frank Lohrberg, Dr.

Professor für Landschaftsarchitektur

Rheinisch-Westfälische Technische Hochschule (RWTH) Aachen/
Deutschland

Büro: lohrberg stadtlandschaftsarchitektur, Stuttgart/ Deutschland

Valerio Morabito, Dr.

Professor für Landschaftsarchitektur

Università degli Studi "Mediterranea" di Reggio Calabria/ Italien

University of Pennsylvania, Philadelphia/ USA

Büro: APscape, Reggio Calabria/ Italien

João Nunes

Professor für Landschaftsarchitektur

Universidade de Lisboa (ULisboa) / Portugal

Università della Svizzera italiana, Mendrisio/ Schweiz

Büro: Estudos e Projectos de Arquitectura Paisagista PROAP,
Lisbon/ Portugal

Jörg Rekittke, Dr.

Professor für Landschaftsarchitektur

Royal Melbourne Institute ofTechnology（RMIT）/ Australien

Robert Schäfer

Fachjournalist Landschaftsarchitektur, Landezine

Ehem. Chefredakteur Garten + Landschaft, TOPOS The
International Review on Landscape Architecture and Urban Design/
München

Frederick Steiner, Dr.

Paley Professor und Dekan der School of Design

University of Pennsylvania, Philadelphia/ USA

Antje Stokman

Professorin für Landschaftsarchitektur

HafenCity University（HCU）, Hamburg/ Deutschland

Dietmar Straub

Professor für Landschaftsarchitektur

University of Manitoba, Winnipeg/ Kanada

Alan Tate, Dr.

Professor für Landschaftsarchitektur

University of Manitoba, Winnipeg/ Kanada

Anna Thurmayr

Professorin für Landschaftsarchitektur

University of Manitoba, Winnipeg/ Kanada

Udo Weilacher, Dr.

Professor für Landschaftsarchitektur und industrielle Landschaft

Technische Universität München/ Deutschland

Christian Werthmann

Professor für Landschaftsarchitektur und Entwerfen

Gottfried Wilhelm Leibniz Universität Hannover/ Deutschland

Kongjian Yu, Dr.

Professor für Landschaftsarchitektur

Architecture, Beijing University / China

Büro: Turenscape, Beijing/ China

晶簇之间

　　意大利风景园林师瓦莱里奥·莫拉比托，于 2007 年出版了他的作品集《纽约风景》[①]。对于他栩栩如生的作品，詹姆斯·科纳给出了这样的评价："莫拉比托的创作，运用了场地速写与注解相结合的手法，是一场奇幻的地理艺术展。这些作品之所以如此令人着迷，不仅仅是因为画面呈现的状态（临摹或描绘的内容），更是因为那些不完美的涂抹痕迹，体现了创作的过程，记录下他是如何观察与思考的"。

　　那时的莫拉比托已经在费城的宾夕法尼亚大学任教。他反复探索着纽约这座城市，把城市看作是景观的特殊形式，用拍摄和计算机绘制的方式去捕捉城市复杂又多孔的空间结构。由此产生了非比寻常的图像分析结果：一种充满人类世特征的风景。图中令人叹为观止的建筑垂直度，令人联想到美国的布莱斯峡谷或中国的黄山。

[①] 英文原著：Morabito, V. *Paesaggio New York*. Biblioteca del Cenide，2007. —— 译者注

　　书中的所有插图均来自瓦莱里奥·莫拉比托，有些是他特意为本书所作。在这些图中，有一种水平方向的力量悄悄地注入垂直生长的纽约世界，仿佛游走在水晶晶簇般的森林之中，它们是如此的鲜活和不受约束。莫拉比托邀请读者跟随他的思路一起去感知和探寻新的可能。

　　向莫拉比托致以诚挚的谢意！

乌多·维拉赫

Zwischen Kristallnadeln

„Morabito's drawings are wonderful geographic expositions in the tradition of field sketches and annotations. They are fascinating not so much for their representational status (what they depict or what they signify), but more for their status as process works, as scribbled traces–inaction that record the labor and discovery of seeing through drawing." So kommentierte James Corner die faszinierenden Zeichnungen des italienischen Landschaftsarchitekten Valerio Morabito, die er in seinem Büchlein *Paesaggio New York* 2007 veröffentlichte.

Damals schon lehrte Morabito an der University of Pennsylvania in Philadelphia und erkundete immer wieder New York City, um fotografierend und am Computer fieberhaft zeichnend, die Stadt als Sonderform von Landschaft mit ihrem komplexen, porösen Raumgefüge zu erfassen. Entstanden sind ungewöhnliche zeichnerische Analysen einer vollkommen anthropogen geprägten Landschaft, deren atemberaubende Vertikalität Erinnerungen an den Bryce Canyon in den USA oder das Huang Shan Gebirge in

China hervorruft.

In den aktuellen Zeichnungen von Valerio Morabito, manche eigens für diese Publikation angefertigt, schleicht sich mitten in die radikale Vertikalität New Yorks eine lineare Horizontalität ein, die scheinbar lebendig, ungebunden und suchend im Wald aus Kristallnadeln umherstreift. Morabito lädt dazu ein, seine Protokolle zu studieren und neuen Pfaden, auch im Denken, zu folgen – herzlichen Dank dafür!

Udo Weilacher

《高线启示》原版于 2018 年 5 月 16 日公开发表，即慕尼黑工业大学建筑学院授予风景园林师詹姆斯·科纳（1961 年出生）荣誉博士学位之时。

Die Publikation erscheint anlässlich der Verleihung der Ehrendoktorwürde durch die Fakultät für Architektur der Technischen Universität München an den Landschaftsarchitekten James Corner（*1961）am 16. Mai 2018 in München.

This publication appeared on the occasion of the conferral of an honorary doctorate on landscape architect James Corner（*1961）through the Faculty of Architecture of the Technical University of Munich on the 16th of May 2018.

著作权合同登记图字：01-2022-6660号

图书在版编目（CIP）数据

高线启示 /（德）乌多·维拉赫（Udo Weilacher）
编著；李梦一欣，黄琦译 . —北京：中国建筑工业出
版社，2022.12
书名原文：Inspiration High Line
ISBN 978-7-112-28245-6

Ⅰ.①高… Ⅱ.①乌… ②李… ③黄… Ⅲ.①园林设
计 – 景观设计 – 研究 Ⅳ.① TU986.2

中国版本图书馆 CIP 数据核字（2022）第 243145 号

责任编辑：李玲洁　段宁
责任校对：李美娜

高线启示
INSPIRATION HIGH LINE
[德] 乌多·维拉赫（Udo Weilacher） 编著
李梦一欣　黄琦　译
*
中国建筑工业出版社出版、发行（北京海淀三里河路9号）
各地新华书店、建筑书店经销
北京雅盈中佳图文设计公司制版
北京盛通印刷股份有限公司印刷
*
开本：880 毫米 × 1230 毫米　1/32　印张：5⅞　字数：126 千字
2023 年 2 月第一版　2023 年 2 月第一次印刷
定价：**49.00 元**
ISBN 978-7-112-28245-6
（40211）